Historical Geography and Geosciences

This book series serves as a broad platform for contributions in the field of Historical Geography and related Geoscience areas. The series welcomes proposals on the history and dynamics of place and space and their influence on past, present and future geographies including historical GIS, cartography and mapping, climatology, climate history, meteorology and atmospheric sciences, environmental geography, hydrology, geology, oceanography, water management, instrumentation, geographical traditions, historical geography of urban areas, settlements and landscapes, historical regional studies, history of geography and historic geographers and geoscientists among other topically related areas and other interdisciplinary approaches. Contributions on past (extreme) weather events or natural disasters including regional and global reanalysis studies also fit into the series.

Publishing a broad portfolio of peer-reviewed scientific books Historical Geography and Geosciences contains research monographs, edited volumes, advanced and undergraduate level textbooks, as well as conference proceedings. This series appeals to scientists, practitioners and students in the fields of geography and history as well as related disciplines, with exceptional titles that are attractive to a popular science audience.

If you are interested in contributing to this book series, please contact the Publisher.

More information about this series at http://www.springer.com/series/15611

Borna Fuerst-Bjeliš · Nikola Glamuzina

The Historical Geography of Croatia

Territorial Change and Cultural Landscapes

Springer

Borna Fuerst-Bjeliš
Faculty of Science, Department of Geography
University of Zagreb
Zagreb, Croatia

Nikola Glamuzina
Faculty of Humanities and Social Sciences, Department of Education
University of Split
Split, Croatia

ISSN 2520-1379 ISSN 2520-1387 (electronic)
Historical Geography and Geosciences
ISBN 978-3-030-68432-7 ISBN 978-3-030-68433-4 (eBook)
https://doi.org/10.1007/978-3-030-68433-4

This Springer imprint is published by the registered company Springer Nature Switzerland AG
The registered company address is: Gewerbestrasse 11, 6330 Cham, Switzerland

Foreword

It is true that geography focuses on current spatial structures. It is, however, also justified to say that physical as well as human geography also look into the past, and refer to the genesis of current spatial and space-related features, as current spatial structures can never be explained just from the modern perspective—they are always the result of several temporal layers. For geomorphology, it is essential to know the factors of feature formation, e.g. whether a feature is of glacial, fluviatile, or other origin. Likewise, human geography delves into the predecessors of current spatial structures, i.e. earlier settlement and transportation networks, central-place systems, administrative and political structures, spatial structures and systems of the economy, socio-economic disparities, earlier stages of urban development, earlier cultural, religious, and linguistic patterns, etc. All this is performed with the background of natural conditions, location, and internal as well as external spatial relations of a given territory.

For human geographers, it is especially interesting to see which older spatial structures have persisted to the present day, and whether earlier structures functioned as a precondition for later ones, as well as which period (s) had the strongest and most lasting impact. In most parts of Europe, the urban system shows remarkable continuity from Roman times; the transportation network almost exactly follows the roads of the eighteenth century, as their pattern was reinforced by the major railway tracks in the nineteenth century. These, in turn, became—with rare exceptions—the blueprint for our modern high-rank road systems, i.e. motorways and highways. To apply this continuity argument to the Croatian case, Kninska Krajina in the northern Dalmatian Hinterland functioned as a border zone along the northern Adriatic coast in the Early Middle Ages for the first Croatian state, and kept this function throughout the Venetian and Austrian periods and on into the contemporary era.

It is, however, also interesting for geographers to learn that sometimes quite the opposite is true that historical discontinuity is also possible and that places have played roles that are difficult to believe today. Thus, there really are no "eternal" peripheries. Socio-economic spatial disparities can change due to endogenic and exogenic forces, e.g. new local initiatives, new currents in the regional or global economy, and the emergence of new high-ranking centres, for which a certain place is more favourably located. The Croatian coast, as well as Pannonian Croatia, are cases in point. Their respective roles, of either centre or periphery, have changed several times throughout history.

Examining spatial structures and their historical development is, however, only one of many currents in human geography, and is not the most vital today. Croatian geography has, nevertheless, a strong tradition in this field. Apart from the *Croatian Historical Atlas* (*Hrvatski povijesni atlas*)—a book about the historical geography of Croatia in maps edited by Krešimir Regan in 2003—it has mainly been represented by Veljko Rogić, to whom the original Croatian edition of this book was devoted and whose works fill several pages in the bibliography. He did not edit any books on Croatia's historical geography but was one of the authors of *The Geography of the Socialist Republic of Croatia* (*Geografija SR Hrvatske*), edited in six volumes in 1974/1975 with a considerable focus on historical geography. Historical aspects also played a significant role in his *Regional Geography of Yugoslavia* (*Regionalna geografija Jugoslavije*) published in 1990. His contribution to *Croatia—regional studies, history, culture, politics, economy, law* (*Kroatien—Landeskunde, Geschichte, Kultur, Politik, Wirtschaft, Recht*), co-edited by the author of this Foreword in 1995, titled *Historical Geography of Croatia* (*Historische Geographie Kroatiens*) can be regarded as a concise and condensed summary of his thoughts on this matter. Therein, he stresses the bridge function of Croatian territories between the Danubian area and the Adriatic.

It can also be taken for granted that this book is, to a larger extent, based on his seminal works and pulls the various streams of his mosaic-like opus into a consistent whole. However, the authors of this book also took it upon themselves to reconsider certain aspects and to regard them from new angles and perspectives, after some time had passed. This is a worthwhile and laudable practice in historiography as well as in historical geography, which can sometimes be overlooked. Less than a quarter of a century has passed since Veljko Rogić's last works, but so much has happened in the meantime —especially to Croatia. Its tourism, and herewith its dominant branch of the economy, has completely recovered—and with it the Adriatic coast as its regional focus; its high-ranking transportation network has significantly been expanded and modernised, essential for territorial cohesion of a small country that is not particularly compact. Additionally, NATO accession in 2009 and EU membership since 2013 have fully integrated Croatia into the political West and the closer community of European countries.

Without intending to summarise or to anticipate the contents of this book, it may be interesting to consider the essentials of Croatian identity and to reflect on the fact that the most significant spatial structures of the country are results of its long and turbulent history.

This is first and foremost its affiliation with the Western cultural sphere. Starting with the first Croatian state in the ninth century, along what is today the northern Croatian coast and into the tenth century expansion into the hinterland, was allied with the Frankish Empire and assumed its Western, Latin version of Christianity as well as its political system that was characterised by a much looser connection between secular and ecclesiastical powers than in the Byzantine sphere. This, in turn, resulted in a softer societal hierarchy and an increased amount of subsidiarity. This led—much in contrast to Greater Moravia—to not accepting the Byzantine mission by Cyril and Methodius and their disciples and to an assumption of a frontier position as a part of the Latin European West, opposite the powers of the East: the Byzantine Empire and, later, the Ottoman Empire (*Antemurale Christianitatis*). This remains a key cultural and political aspect of Croatia to the contemporary era, and will certainly continue to be so for the foreseeable future.

The second marker of Croatian identity, closely connected with the first, is Catholicism. Contrary even to Slovenes and Poles, Croatian lands were never a fertile soil for the Reformation. Comparable only to the situation in Poland, Slovakia, and Malta, the role played by the Catholic Church in contemporary Croatian society is among the strongest of any European country.

The third, very geographical characteristic of Croatia, is its cultural divide into Mediterranean and continental, i.e. Pannonian, parts. This is certainly supported by natural conditions, but was first established in Roman times, when the urban network was much denser and the countryside was much more advanced and open in terms of agriculture, especially along the coast. This divide was essentially accentuated by Byzantine, and later Venetian (the Republic of Ragusa can be regarded as Venetian in the cultural sense, despite never being under Venetian control) and Austrian political domination of the coast, while Pannonian Croatian lands had to accept Hungarian domination of varying intensities for the greater part of their history. This meant a strong impact of Roman Mediterranean culture on the coast, of which Venice had the largest impact, perpetuated under Austrian rule and is still apparent in the cultural landscape. In contrast, there was a Central European transformation and reorientation of Croatia's Pannonian areas.

Beyond this principal cultural divide, however, Croatia is composed of historical–cultural landscapes with strong identities. This is a heritage of the political division of Croatian lands, from the first Croatian state in the Early Middle Ages to socialist Yugoslavia, prior to Croatia becoming a political unit in its current shape for the first time. The so-called Triune Kingdom of Croatia, Slavonia and Dalmatia was never a political reality, and the ephemeral Ustaša state of World War II did not include Istria or even all of Dalmatia. Thus, landscapes like Istria, Dalmatia, Slavonia, Srijem, Hrvatsko Zagorje, and at a lower level also Banovina, Kordun or Lika are—due to their specific history and cultural formation—deeply rooted in the consciousness of their respective population. This still affects discussions on administrative subdivision and regionalisation at the time of writing and occasionally prompts regionalist movements that strive for greater autonomy.

As the final aspect of Croatian identity and the nation's self-image is its status as "motherland" for a sizable diaspora that is spread throughout the world. This diaspora adds to the total number of Croats in the world and comprises not only the rather small autochthonous Croatian minorities in Austria, Hungary, Slovakia (earlier also in Czechia), Romania, Montenegro, and Italy as well as the larger ones in Bosnia and Herzegovina and Serbia, but also the emigration waves overseas in the late-nineteenth century up to World War I (mainly from Dalmatia), as well as to Western Europe in the Communist period and later. There is a great deal of literature regarding emigration, and emigration is an important topic in local cultures of remembrance.

As it is essential for geographers to be aware of the historical formation of cultures and landscapes, this book provides an excellent introduction into these aspects of a country (in this case, Croatia), which are—due to its vivid history—of special importance.

February 2020

Peter Jordan, Ph.D.
Honorary and Associate Professor
Institute for Urban and Regional Research
Austrian Academy of Sciences
Vienna, Austria

Faculty of Humanities
University of the Free State
Bloemfontein, South Africa

Preface

The idea to write a book on the historical geography of Croatia goes back almost 10 years, when we (professors of historical geography at the University of Zagreb and University of Split) were discussing the need to develop a university-level textbook for our students. Until that point, we had all been using a classic text from the twentieth century, written by our Prof. Veljko Rogić—a doyen of historical geography of Croatia. Due to the turbulent changes in geography that wracked Croatia following its independence, it became clear that this classic text needed an update. When we started to work on the textbook, we were extremely lucky to be able to take advantage of Prof. Rogić's broad knowledge and experience, as well as his insightful and pointed critiques. The book was published on his 90th birthday.

After some time had passed, we noticed that interest for this text had gone beyond the boundaries of the core disciplines of geography and history, as well as the boundaries of its Croatian readership. There was constant interest in the book evident on research and academic networks, but it was inaccessible to those who didn't speak Croatian. Moreover, there are no other contemporary books, in Croatian or any other language, that cover the historical geography of Croatia, making the topic generally inaccessible to a great number of potential readers.

When Springer launched a new book series on Historical Geography and Geosciences, we came up with the idea to publish an English edition of the book. The Springer publishing editor of the book series at the time, Dr. Michael Leuchner, has since helped us enormously with making the book project a reality.

This book, however, differs from the original Croatian textbook version. As a historical–geographical monograph of Croatia, this new version has been thoroughly revised and adapted for the wider interested readership. This edition is aimed both at scholars who have previous knowledge of the subject matter as well as those who are new to Croatia's history and/or geography.

The book introduces the reader to a broad and thorough political and socio-cultural framework that shows how the space and cultural landscape of Croatian territory were shaped, for every respective period of its historical–geographical development. The book is structured according to nine main historical–geographical periods, from prehistory to the present day. The introductory chapter gives an insight into the territory of Croatia in terms of its location and position in geographical, historical, political, cultural, and environmental terms, which is essential for further understanding of the area's historical–geographical conditions and development trajectories. The last chapter, in contrast, relates to the post-socialist period (1991–2019), which is too short to identify a substantial cultural landscape change. Although this chapter represents a step out of the domain of historical geography, it aims to identify the basic processes of the post-socialist and post-war period that have initiated geographically relevant trends and cultural landscape change.

The book includes thematic maps and photographs. The choice of introductory photographs for each chapter is to show the distinctive cultural landscape or architectural heritage from different times and regions of Croatia, which illustrate the most characteristic features of each respective historical–geographical period. Adriatic, continental, mountainous, and Pannonian Croatia (areas that significantly differ in cultural landscape features, traditions, and historical and cultural development) are equally represented by these photos.

Many people and organisations have contributed in bringing this book to fruition. We are indebted to Prof. Rogić, who was much more than our professor; he opened the window to the beautiful world of historical geography for us. Additionally, we cannot adequately express our gratitude to Jadranka Lady Beresford-Peirse, founder and trustee of the *International Trust for Croatian Monuments*, without whose understanding and generous support this book project would never have been possible. We are also cordially thankful to Fra Branko Brnas and the Zaostrog Monastery for their trust in the value of the project and support for translation and editing. We would also like to offer our sincere gratitude to the Split-Dalmatia County as well as to the University of Zagreb (*Research Grant no. 108-F19-00010Palaeoenvironmental changes of Croatian Karst*) for supporting the translation and cartography expenses.

We would like to extend our thanks to Prof. Marko Trogrlić, who has been a constant source of support for us starting from the inception of the Croatian edition. We also would like to thank Prof. Jurica Matijević and Biserka Cimeša for their encouragement, and valuable advice and suggestions.

We are also very thankful to our librarians (some of the most important people for scholars and researchers): Ljiljana Bajs from the Central Geographic Library at the Department of Geography, Faculty of Science, University of Zagreb; Snježana Dimzov and Nataša Torlak from the Library of the Faculty of Humanities; and Vlasta Ugrinović and Ivana Ljubić from the Faculty of Science of the University of Split.

We owe particular gratitude to all the people and institutions who have given us the permission to freely reproduce their photographs (in alphabetical order): Barbara Bjeliš, Ivana Erdelez, Siniša Sakoman, Ivan Zupanc, the Šibenik City Museum, and the Croatian National Tourist Board. We are also very thankful to the Miroslav Krleža Institute of Lexicography for their kind permission to use their cartographic material and sources. Concepts and content of the maps and other figures are authored by the authors of the book and were drawn by *Srednja Europa d.o.o.* and Ivica Rendulić and hereby we express our thanks for their cooperation.

The translation itself was very demanding work. The problem of how to deal with original specific terms or names, especially those from distant past, required constant communication and cooperation with the translator. This was especially important, because in many areas there were no established works or widely accepted terminology to lean on. The basic starting point was to respect Croatian terms or names as far as they could still be easily understood by readers who are not terribly familiar with Croatia or the Croatian language (more on this in the translator's note). We are so very thankful to our translator, Christopher Grey Kaufmann, for his close cooperation throughout the process and his valuable opinions and suggestions. He also discretely warmed up the sometimes very strict and cold scholarly text with numerous tiny and soft touches, which we appreciate very much.

Finally, we are indebted to the entire team at Springer during the time of the book production, to Michael Leuchner, Doris Bleier, Carmen Spelbos, Dewika Lachman, and Ambrose Berkumans for their constant support in bringing this book to fruition.

Zagreb, Croatia
Split, Croatia
February 2020

Borna Fuerst-Bjeliš
Nikola Glamuzina

Translator's Note

The English text deviates somewhat from the original Croatian version in order to be more accessible to an audience that doesn't speak Croatian or necessarily know much about the country and its people. This is not only a result of the translation, but of careful choices in terms of terminology and what background knowledge is assumed on the part of the reader. Rather than automatically defaulting to international terms (for events, places, treaties, etc.), Croatian terminology has largely been used. Additionally, certain sections have been mildly altered in order to ensure that context is provided. For readers interested in how to pronounce the Croatian terminology found in the book, we have included a short guide below.[1]

Croatian differs from English in that it is a phonetic language, meaning all letters get pronounced and aren't modified by neighbouring letters.

Pronouncing the Croatian letters: č, ć, dž, đ, š, ž, lj, and nj (and j and r)

Croatian has a few letters that contain two characters but are considered to be a single letter and are pronounced as a single sound. These are **dž**, **lj**, and **nj**.
nj is pronounced like ni in "minion".
lj sounds like lli in "billion"
dž sounds like dg in "wedge".
The rest are single characters that represent sounds that are often achieved using 2 (or more) letters in English. These are **č**, **ć**, **s**, **š**, and **ž**.
č is pronounced like tch in "catch".
ć sounds like the ch in "chew" (this letter is similar to **č** but is pronounced more softly).
đ sounds like the g in "general" (this letter is similar to **dž** but is pronounced more softly).
š is pronounced like the sh in "shoe".
ž is pronounced like si in "illusion".
While **j** and **r** are present in English, Croatian uses them rather differently.
j sounds like the y in "yes".

[1]For interested readers, here is a more complete guide on Croatian pronunciation: https://www.learncroatian.eu/blog/croatian-pronunciation-basics

r is pronounced strongly (many Scottish accents pronounce **r** similarly) and sometimes functions as a vowel—Croatian words often string alarming amounts of consonants together and some words even lack vowels entirely (e.g. *smrt* and *čvrst*).

Zagreb, Croatia
February 2020

Christopher Grey Kaufmann

Contents

Introduction—The Territory

1

Abstract

The chapter discusses the basics of the territory of Croatia, which are of importance for its distinctive historical–geographical development. Its location and especially its position in geographical, geopolitical/geostrategic, and cultural terms is simultaneously the result and cause of its historical–geographical development. Foreign influences came from the west and southwest (Mediterranean), as well as from the north and northwest (Central Europe). Mediterranean influences have been present since Antiquity, while Central European influences began in the Early Middle Ages and were greatly strengthened after the suppression of Ottoman state organisations. Because both Mediterranean and Central European cultural influences spread by way of strong social organisations, their traces have remained visible in various forms (urbanism and architecture, art, dialects, mentality, gastronomy, music, and individual elements of folk culture) in individual parts of contemporary Croatia. Furthermore, the chapter reviews the process of bordering in Croatia. The contemporary territory of the Republic of Croatia was determined during the Second World War and was precisely geographically and cadastrally defined shortly before the war ended. The chapter also focuses on the natural environment, which is very diverse, despite the country's relatively small size. Considering the basic elements of the natural environment—relief, climate, soil, and vegetation—there are three different, important environmental wholes: Pannonian, Dinaric, and Adriatic/Mediterranean. Throughout history, each type of natural environment determined not only the manner of valuation of land, but also how humans altered the environment, shaped cultural landscapes, and organised territories.

Keywords

Croatia · Geographical position · Mediterranean · Central Europe · Balkans · Borders · Natural environment · Dinaric · Pannonian · Adriatic · Landforms · Climate

The temporal dimension in geographical research has been present since the beginning of geography itself. It is a fact that current geographical characteristics and trends cannot be explained without analysis of past processes. Historical geography is based on the historical dimension of observation of the Earth's surface. Therefore, we can define it as the geography of past eras, i.e. as a geographical discipline in terms of methodology and research focus.

Historical geography, however, is often understood as a bridge between geography and history—fields that occupy different intellectual fields, but have many areas of overlapping

B. Fuerst-Bjeliš and N. Glamuzina, *The Historical Geography of Croatia*,
Historical Geography and Geosciences, https://doi.org/10.1007/978-3-030-68433-4_1

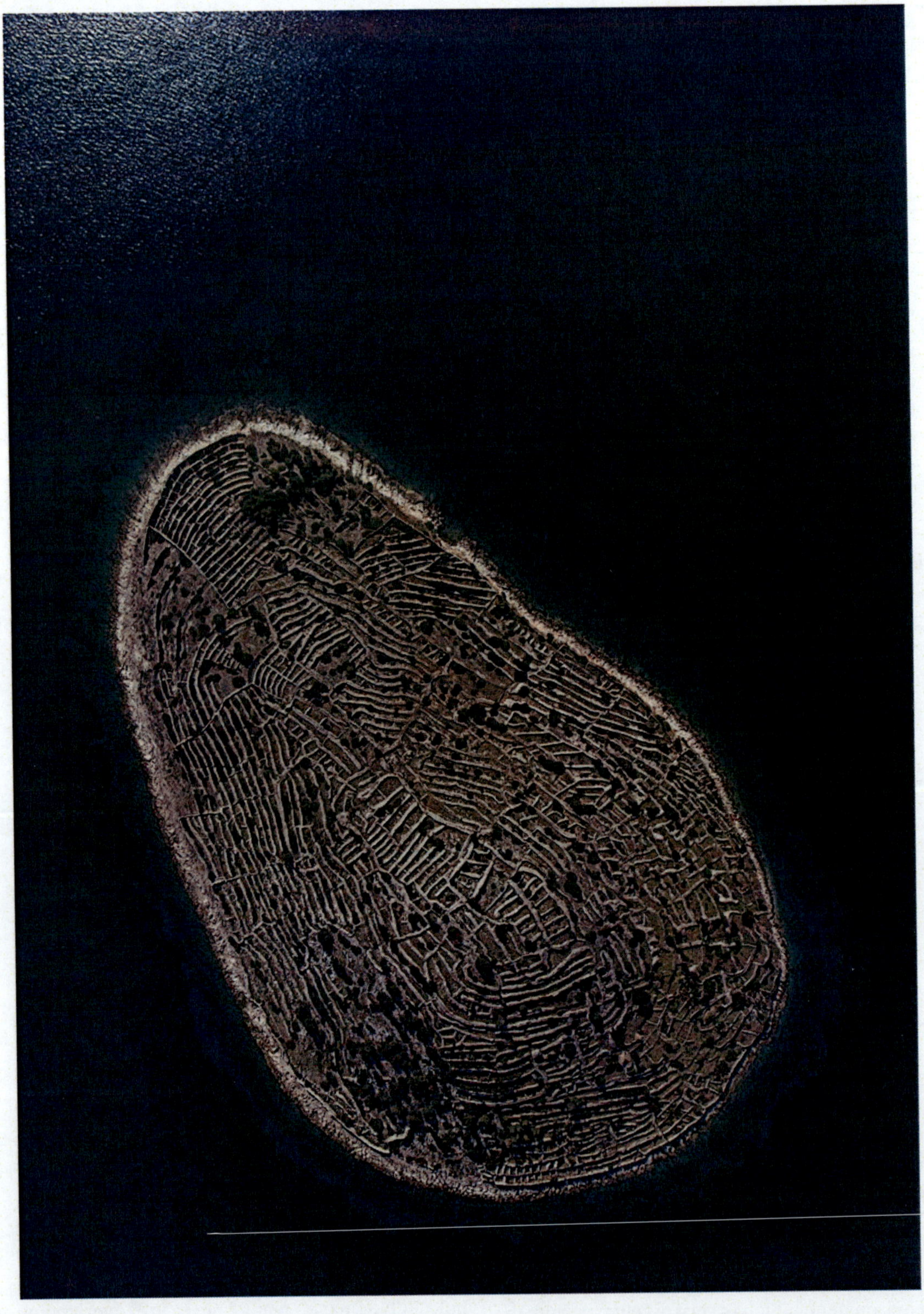

Baljenac: an island in the Šibenik Archipelago, Dalmatia The networks of stone walls, traditional elements of the cultural landscape of Croatian karst areas, are evidence of millennia of human influence on the environment. The lack of workable soil demanded their construction. This served to create and protect workable land by removing the stones and building short walls. Baljenac island was covered by vineyards during the great wine conjuncture (Photograph: Matko Jakelić. Reproduced with permission of the Šibenik City Museum)

research. Geography and history are different ways of looking at the world that are essentially linked. In the past, historical geography often had the status of sub-discipline between history and geography. With the strengthening of interdisciplinarity and cooperation among scientific disciplines, however, new areas of research and perspectives for historical geography have opened (Butlin 1993; Schenk 2011). Historical geography gives a unique contribution to new scientific, interdisciplinary studies, especially of environmental history (Williams 1994), which has in turn opened new areas in which geographers and historians can collaborate (Hughes 2011).

According to the nature of the discipline, the historical geography of Croatia can be understood as the geography of Croatia's past. It is primarily the geography of how the space and cultural landscape have been shaped, for every respective period of Croatia's historical–geographical development. To determine the substance, meaning, and nature of the processes underpinning the space and cultural landscape of Croatia, it is necessary to provide a broad temporal and spatial context, i.e. a thorough political and socio-cultural framework for each era. The basic approach to historical geography of Croatia that we took was to analyse spatial and cultural landscape development and change, embedded in the socio-cultural context of each respective era.

In order to understand historical–geographical conditions and trajectories of development, the introductory chapter of the book provides broad insight into the territory of Croatia in terms of its position (physical, cultural, religious, etc.) within Europe, the process of bordering, and the general environmental setting.

1.1 Croatia—Territory and the Process of Bordering

Croatia is a European country with a total area of 88,073 km^2. Its territory stretches between longitude 13^{0}29′ and 19^{0}26′ E, from the Istrian Peninsula in the west to Srijem (Syrmia, Sirmium) in the east, and between latitude 46^0 33′ and 42^0 22′ N, from the Međimurje Region to southern Dalmatia.[1] Of the country's total area, 56,594 km^2 is made up of land territory while the remaining 31,479 km^2 consists of interior sea waters and territorial seas. Croatia has 4,174,349 inhabitants (2016).[2] Despite its relatively small area, Croatia consists of very diverse natural areas—largely Pannonian, Mediterranean, and Dinaric—due to the characteristic bow-like shape of its territory (Fig. 1.1).

1.1.1 The Position of Croatia Within Europe and the Mediterranean

The geographical position of a given territory includes a litany of socio-geographical, cultural, economic, and historical–geographical elements that are, in the case of Croatia, significantly more important than its size in terms of area and population. Each of the aforementioned three types of geographical areas encourages a specific manner of human life, which can be observed in the corresponding tangible and intangible elements of the cultural landscape (e.g. dialects, regional consciousness, and mentality). Consequently, three different socio-cultural zones have been shaped throughout history in contemporary Croatian territory. These are the old Mediterranean agrarian area, the younger and less-developed Pannonian and peri-Pannonian agrarian areas, and the traditionally pastoral Dinaric area. The Dinaric region is also an area where other cultural influences (specifically, of the Mediterranean and (peri-)Pannonian regions) have become interwoven (Rogić 1971b, 1976a).

Due to this, the Dinaric area is often called a transitory area, and its similarity to other Dinaric areas in neighbouring countries is often stressed (Crljenko 2014). The peripheral areas of Croatia

[1]Statistical Yearbook 2017, Croatian Bureau of Statistics.

[2]First release, Year LIV, 14 Sept 2017, Croatian Bureau of Statistics.

Fig. 1.1 The Republic of Croatia within its current borders

within the Mediterranean and (peri-)Pannonian areas have been heavily exposed to foreign cultural influences throughout history, as opposed to the less-accessible Dinaric mountainous spaces that retained components of their traditional indigenous culture longer (Rogić 1976b). Throughout history, this geographical diversity marked the distinction between two ancient types of economies that, in turn, defined respective ways of life and with them related social structures and mental contexts. First, there is the traditional pastoral, primarily transhumant, type of economy that determined the shaping of mountain tribal societal structures typical for the Dinaric area. The second dominant agrarian type of economy is defined by a stabilised sedentary type of settlement and advanced forms of societal structures and political organisations that are characterised by the permeation of foreign cultural elements (Rogić 1972, 1982).

Foreign influences came from the west and southwest (Mediterranean), as well as from the north and northwest (Central Europe). Mediterranean influences have been present since Antiquity, by which Roman (primarily Italian) and Catholic cultural influence spread along the Croatian coast. Central European influences began in the Early Middle Ages, and were greatly strengthened after the suppression of Ottoman state organisations. The bearers of this influence were Germanic peoples of the German cultural group, and also Hungarians starting in the Late Middle Ages. Being that both the Mediterranean and Central European cultural influences spread by way of strong social organisations, their traces have remained visible in various forms (urbanism and architecture, art, dialect, mentality, gastronomy, music, and individual elements of folk culture) in individual parts of contemporary Croatia. Cultural influences from the internal parts of Southeastern Europe, which came from the Dinaric area during the era of Ottoman conquest and expansion, are less evident. In this way, elements of the military frontier way of life, based on rural mountain clan organisations typical for Dinaric pastoral areas, as well as Orthodox Christian and Islamic cultural influences, spread (Crljenko 2014).

Because Croatia belongs to the peripheral part of Southeastern Europe from a geographical viewpoint, bordering both the Mediterranean Sea and Central Europe, its "bridge" role in linking the Mediterranean with Central Europe has also given great strategic significance to Croatian space throughout history. Due to this fact, it has been the scene of turbulent political events throughout history.

The contemporary geographical position of Croatia is inseparable from the transitory–geographical meaning that Croatian territory has had throughout history, and which is especially evident in today's modern transportation era. This arises from the fact that the Adriatic Sea is the part of the Mediterranean that extends the furthest into continental Europe, and is therefore closest to Central Europe geographically. Additionally, the Pannonian Basin is closest to the Mediterranean in the area of Croatia. The great transitory and geographical importance of the Croatian area, already eyed greedily in Antiquity by the ancient Romans, arises from this fact; and the systematic valorisation of the region started under Habsburg administration (from the beginning of the eighteenth century). This transitory area is the narrowest on a line stretching from the southeastern edge of the Pannonian Basin (Karlovac) to Kvarner (Rijeka, Bakar, Senj), which is often called the "Croatian Corridor" in scholarly literature, i.e. the shortest and narrowest transverse passage between the southeastern edge of the Pannonian Basin and the northern Adriatic.

The greatest impediment of the relief in the transit corridor lies in the area of Gorski Kotar, a relatively low part of the Dinarides that is easily traversed. The average elevation (from 700 to 900 m a.s.l.) and development of the river valleys and mountain passes have enabled relatively easy transitory linkage (Rogić 1971a, 1975). The Croatian Corridor, along with its military-strategic meaning, still has international significance today, as it has the important role of linking the transit between the Pannonian Basin and the Mediterranean.

The geographical position of Croatia itself is sometimes also defined by the concepts of Central Europe and the Balkans. The concept of Central Europe denotes the region between the Atlantic parts of Europe of expressed maritime character and the continental region of Europe. The problem in this definition of Central Europe can be found in the fact that it is a region without expressed geographical boundaries. Therefore, the region of Central Europe can be understood in two ways: in the narrower or broader sense of the term. The narrower definition is understood to mean Germanic Central Europe which encompasses three different spatial wholes: the northern plains, the highlands, and the Alps—a region of the European continent in which German ethnic and cultural elements prevail and enjoy the main political and economic roles.

The broader definition is understood to mean a Germanic–Slavic–Hungarian region which also even includes some individual bits of the Romanian region. Defined as such, Central Europe additionally encompasses other large natural wholes: the Polish lowlands, the western peripheral parts of the Eastern European Plain, the Czech Basin, the Carpathian Mountains, the Pannonian Basin, and the edge of the Wallachian Plain. This sort of conception of Central Europe is tightly linked with the geo-political concept of the influential region of Germany as a great power (Rogić 1990), characterised during the turn from the nineteenth to the twentieth century, and is often described by the German concept Mitteleuropa. Following the end of the Second World War, the broader concept of Central Europe ceased to exist as a geographical reality due to the fall of the "Iron Curtain" that divided the Continent into two wholes: Eastern Europe (in the context of the former communist bloc) and Western Europe (dominated by western capitalist countries). Germany itself was divided into two states in this era, each with its own pervasive socio-political system. Thereby, the broader concept of Central Europe disappeared during the Cold War, but has slowly come back into use after the end of the Cold War and the reunification of Germany, which has returned to its position as the key political and economic force in Continental Europe. Accordingly, the problem with categorising Croatia into this group of Central European states lies in the differing understandings of the very concept of Central Europe, as well as in the fact that this is not only a region in the geographical sense of the word, rather an expressly heterogeneous region (from a natural and social viewpoint) for which the only connection is the strong political and economic influence of Germany.

In contrast, in neighbouring countries, as well as in other parts of Europe, Croatia is often categorised into the part of Europe referred to as the Balkans. The fact that this concept is more often used in other parts of Europe should come as no surprise as the term does not even originate from the part of the continent for which it is used. The very concept of the Balkan Peninsula (Balkans)[3] was first used by the German geographer August Zeune in 1808, primarily to describe the part of Europe that had been under Ottoman control. Namely, this part of the continent was marked by major differences from a socio-economic viewpoint, as well as in cultural development related to the neighbouring northern area within the Habsburg state organisation, and it has since become generally linked with concepts of backwardness and primitivity.

The problem, however, is that the Balkans is an imprecise and illogical concept in a geographical sense. On the one hand, it is an area that is hard to call a peninsula as it doesn't have an expressed peninsular shape, while maritime characteristics, which are most reflected in climate characteristics, are very weakly expressed and then only in the peripheral areas. Furthermore, another problem also lies in the borders of this alleged "Peninsula". Namely, one defining characteristic for all large (and small, for that matter) peninsulas is that they taper markedly where they meet the main continental landmass. This can be easily observed in other European peninsulas, e.g. Iberian, Apennine, Scandinavian, Cimbrean, and Crimea. Nevertheless, in the so-called Balkan Peninsula the situation is the exact opposite because, as opposed to tapering, it is the widest part that is in contact with the main European continental mass. From this also stems the question of its northern boundary, which different authors have marked using different natural features—the Danube and Sava rivers to the Bay of Ljubljana (Jovan Cvijić), the Sava and Krka rivers, the Postojna Gate, and further to the Vipava and Soča rivers to the Bay of Trieste (as defined in the Croatian Encyclopaedia—Hrvatska Enciklopedija—from 1941), to the most-used border markings of the Danube, Sava, and Kupa rivers to the point where the Bay of Rijeka penetrates farthest into the mainland.

It is interesting that August Zeune himself, the originator of the concept of the Balkan Peninsula, never defined its northern border (Roglić 1950, 1990). Accordingly, this means that this is an area that, from a geographical point of view, can not in any sense be called a peninsula. It is worth mentioning that the Balkan Peninsula as a concept does not stem primarily from geographical

[3]The term "Balkan Peninsula" is sometimes used exclusively as a "geographical" concept, but the term "Balkan" or "Balkans" is used in sociological, ethnological, political, and other related contexts. The name itself comes from the name of a mountain range in Bulgaria and Southern Serbia.

characteristics, rather from socio-political characteristics based on political and social realities of the era when the concept came into being. Today, however, the continued use of the term ignores the fact that during the last three centuries, and especially during the last hundred years, the socio-economic gap between the regions once under Ottoman control has significantly decreased in relation to their northern neighbours. Therefore, both the cultural and socio-economic significance of the concept of the Balkans have largely lost the meaning they once had.

1.1.2 The Formation of Contemporary Croatian Borders

The contemporary territory of the Republic of Croatia was determined during the Second World War by the decisions of the State Anti-fascist Council for the National Liberation of Croatia (ZAVNOH), confirmed in the documentation of the Anti-Fascist Council for the National Liberation of Yugoslavia (AVNOJ) from 1943, and was precisely geographically and cadastrally defined shortly before the war ended. It is important to emphasise how, during the era of socialist Yugoslavia, the borders between Yugoslavian federal units on land were clearly defined, but the sea borders were not. Contemporary disputes between Croatia and the other three former Yugoslav republics with access to the Adriatic Sea (Bosnia and Herzegovina, Montenegro, and Slovenia) are a result of this.

The total length of the Croatian border is 2,370.5 km, which is very long for a country of Croatia's size. Its length is a result of the (bow or boomerang-like) shape of the nation's territory. The longest section of the border is shared with Bosnia and Herzegovina (1,010.9 km), followed by Slovenia (659.1 km), Hungary (355.2 km), Serbia (325.8 km), and finally Montenegro (19.5 km).[4] Croatia has a border with Italy but only on the sea, and the territorial waters between these two countries come into contact in the waters around Vis and the northern Adriatic.

The greater part of the border with Bosnia and Herzegovina was formed during the era after 1699, when the Treaty of Srijemski Karlovci (Karlowitz) was signed, to the Belgrade Peace Treaty in 1739, with minor corrections following the Second World War. The final correction was made in 1952, when 5 settlements from what was then the territory of Bosnia and Herzegovina were annexed to Donji Lapac, which had been on the border up to that point (Boban 1995). The demarcation between Croatia and Bosnia and Herzegovina in the area of the Neum Corridor, a 24.5 km strip of land by which Bosnia and Herzegovina access the Adriatic Sea, cuts a large part of southern Dalmatia off from the rest of Croatian territory. The Neum Corridor itself also took its final shape during and shortly after the end of the Great Turkish War (1683–1699) at the insistence of the Republic of Ragusa (Dubrovnik).

Namely, the leaders of Dubrovnik wanted to avoid having a land border with the Republic of Venice, due to the territorial ambitions of the latter. During the Great Turkish War, because the Venetian army had taken Dubrovnik's hinterland and completely surrounded the small republic in 1694, the leaders of Dubrovnik pushed for the creation of the Neum Corridor along their northern border and the Sutorina Corridor in the south, as part of the peace treaty that would end the war. Both corridors belonged to the Bosnian Eyalet of the Ottoman Empire, and were confirmed in the later peace treaties of Požarevac (Passarowitz 1718), Belgrade (1739), and Svištov (Sistova 1791). Croatia's land territory is bisected by the Neum Corridor, which was retained after the Second World War as Bosnia and Herzegovina's only access point to the sea.[5]

The border with Slovenia formed spontaneously in the Middle Ages as an ethnic and political boundary, and individual parts of the

[4]Statistical Yearbook 2017, Croatian Bureau of Statistics.

[5]The other point of sea access for Bosnia and Herzegovina (Sutorina Corridor) was annexed by Montenegro in the same era, and the criteria by which the leaders of the time made these decisions remains unclear.

border were later defined under the Austro-Hungarian Monarchy and divided between Croatia and Slavonia on one side, and Carniola and Styria on the other. The Croatia–Slovenia border was in greater part defined after the Second World War, and adjusted in 1954 after Yugoslavia annexed Zone B,[6] which consisted of the contemporary Slovenian coast and the northwestern part of Istria. At the same time, Italy annexed Zone A, consisting of Trieste and its immediate surroundings. In the same year, the border at Žumberak was also defined. The border was finally confirmed with minor corrections in 1956, when Slovenia gained eight additional Istrian villages, which had been within Croatian territory until that point (Klemenčić 1991).

The border with Hungary on the Mura and Drava rivers is the oldest of Croatia's natural borders, and came into use in the tenth century based on the ethnic principle. This border was precisely determined by the peace treaty in 1920, when the possession of the Međimurje Region and three settlements on the left side of the Drava River (Gola, Ždala, and Repeš) in the west, and Baranja (Baranya) Region in the east was confirmed as a part of the Kingdom of Serbs, Croats, and Slovenes (SHS).[7]

Nevertheless, it is important to note how the contemporary borderline doesn't hold exactly to the Mura River, rather it meanders from one side of the river to the other. This is even more evident along the Drava, the bed of which is largely within Croatian territory, as well as individual parts of the left bank.[8]

Today, the border crosses the flow of the Drava only in a few places in Slavonian Podravina, and some parts of the right bank belong to Hungary. The border was defined in the field in Baranja by an international committee in 1922, thereby splitting the region between two states. According to the last pre-war (WWI) Austro-Hungarian census from 1910, in the territory of contemporary Croatian Baranja, Hungarian was the spoken language (but not necessarily mother tongue) of 39.63% of inhabitants, and the second largest share was of German (27.45% of inhabitants). Despite this, the southern part of Baranja was annexed by the Kingdom of Serbs, Croats, and Slovenes (SHS) for economic and strategic reasons. Namely, the area gravitated toward Osijek in economic terms, and it had already been decided that the city would belong to the newly founded Yugoslavian state.

Osijek was crucial in terms of control over Baranja, due to its location on the raised Slavonian Pleistocene terrace in the area where it is closest to the Baranja Pleistocene terrace. Due to its position, Osijek ensured the shortest and easiest passage from Slavonia to Baranja, making it a prime location to rule and oversee, not to mention militarily threaten, the southern part of Baranja. This fact has been confirmed by historical events from Antiquity through the Hungarian Revolution of 1848 (Bösendorfer 2010).

Croatia's contemporary border with Serbia (Voivodina) was formed after the Second World War and is largely natural, as it primarily (but not completely) follows the Danube and its tributaries. As is the case with the Drava, the border jumps back and forth across the river, such that

[6]The peace treaty with Italy signed in Paris in 1947 created the Free Territory of Trieste, to be guaranteed by the United Nations Security Council. It was divided temporarily into northern and southern zones: Zone A, which included the city and its northern hinterland, was put under U.S./British military administration; and Zone B, to the south of the city, was put under Yugoslav administration. The London Agreement of 1954 granted Zone B and part of Zone A to Yugoslavia (523 square km) and the city of Trieste and the remainder of Zone A (236 square km) to Italy. With the signing of the Osimo Treaty in 1975, Italy relinquished all claims to Zone B, finally settling the status of the region and firmly defining the border between Italy and Yugoslavia.

[7]The Kingdom of Serbs, Croats, and Slovenes (SHS) was declared on January 12, 1918. It came into being with the unification of the State of Slovenes, Croats, and Serbs (in the area that had once belonged to Austria–Hungary) and the Kingdom of Serbia and Montenegro. The name "Kingdom of Yugoslavia" was officially ushered in 1929, replacing the former name of the Kingdom.

[8]The greatest part of the left shore of the Drava, which belongs to Croatia, is found in Podravina, following a line from Đelekovec to Novo Virje. The notions of "left" or "right" banks of the river are not subjective, rather they refer to the position of a given river's banks if one is facing the direction of the river's flow, i.e. from its source towards its conflux.

individual parts of the left shore of the Danube[9] belong to Croatia, and a few small parts of the right shore in Baranja are part of Serbia's territory. Just before the end of the Second World War in June 1945, a 5-member committee was named, led by Milovan Đilas—a Montenegrin politician who was one of the leaders of the Yugoslavian Communist Party. After surveying the situation on the ground, the Committee recommended what would become the current border between Croatia and Voivodina, whereby the majority of Srijem—with the exception of its western part—would belong to Voivodina and Baranja would belong to Croatia, according to the ethnic principle. The border with Montenegro relied on the southern border of the Republic of Ragusa (Dubrovnik) (already determined in the fifteenth century), whereby part of Dalmatia (the Bay of Kotor) and the Sutorina Corridor (Bosnia and Herzegovina's southern access to the sea) were annexed by Montenegro.

Croatia's contemporary border with Montenegro is based on the borders of the Dubrovnik administrative area "oblast" (according to the administrative division of the Kingdom of Serbs, Croats, and Slovenes of 1922). The sea border with Italy was determined with a Treaty of Osimo (1975) when the Adriatic Sea border was negotiated between Italy and Yugoslavia (Boban 1995).

1.2 Outline of the Environmental Setting

The natural environment of Croatia, despite the country's relatively small size, is very diverse. Considering the basic elements of natural environment—relief, climate, soil, and vegetation—there are three different, important natural wholes: Pannonian, Dinaric, and Adriatic/Mediterranean. Throughout history, each type of natural environment determined a specific manner of life for the humans living in it—not only in the manner of valuation of land, but also in how humans altered their environment, shaped characteristic cultural landscapes, and organised territories. For example, in the earliest history, the location of a water source in the water-scarce karst was a key element of the first forms of the spatial organisation of life, i.e. the first territorial unit: *comitatus* (*župa* in Croatian). The manner of land parcelling, i.e. distribution of the most valuable arable land in an area where there wasn't much to go around (as was the case on Croatian Mediterranean karst islands), has been preserved since Antiquity to present day in Stari Grad on the island of Hvar—which is protected as a UNESCO World Heritage cultural landscape. Moreover, the fact that the lowest and narrowest parts of the Dinarides are found in the area of Croatia, in the so-called Croatian Corridor, determined the transit–economic orientation of the wider Central European gravitational area toward the Adriatic, i.e. toward the Mediterranean through the territory of Croatia itself. This gives the area considerable strategic value.

1.2.1 Landforms

Croatia consists of three main geomorphological regions: the Pannonian Basin, the Dinarides, and the Adriatic—which geomorphologically mostly refers to submarine relief. The Dinaric region (except for a few isolated spots in the Pannonian region) is also a region of karst that occupies approximately almost half of the territory of Croatia (Fig. 1.2).

Croatia belongs to a temperate fluvio-erosional zone, dominated by fluvial processes in the north and karst and fluviokarst processes in the south. Complex tectonic evolution is coupled with the influence of different exogenic processes (slope, fluvial, abrasional, glacial, periglacial, karst, fluviokarst, and aeolian processes) in various zones of elevation (Bognar et al. 2012).

1.2.1.1 The Pannonian Basin

The Pannonian Basin is a prevailingly lowland subsidence area dominated by large *alluvial fans*, wide *floodplains,* and fluvial *terrace* plains,

[9]This refers to the left shore of the Danube up and downstream of Batina, as well as the area of Kopački Rit and Mohovo.

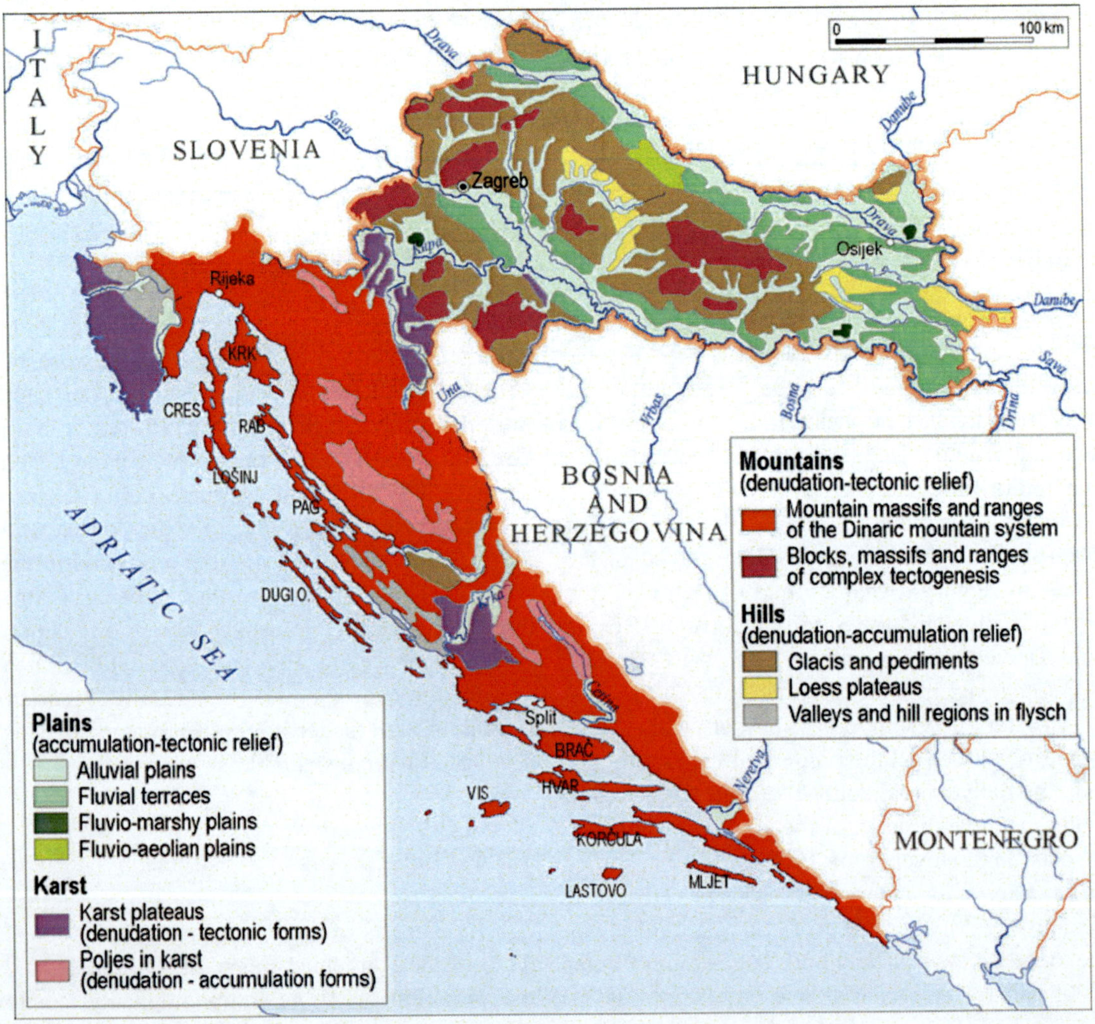

Fig. 1.2 Relief categories in Croatia (Modified after Bognar 1995a)

which are often additionally elevated by thick deposits of loess and loess-like sediments (Bognar 1985a). Pannonian Croatia is a region with prevailing surface outflow, contrary to the Dinaric karst region with karst hydrology where, in spite of abundant precipitation, most of the water disappears underground because of fractured carbonate complexes. The northwest part of the Pannonian Basin has the greatest concentration of surface water and the highest drainage density. The Sava River is the hydrogeographical axis draining the greater part of the area (Fig. 1.3).

In almost all floodplains, one can distinguish higher and lower levels associated with enduring flooding and/or low water. Rivers have no marked valleys, only fluvial plains. Their properties are attributed to neotectonics and to recent tectonics as well as to regulation and land reclamation works dating from the eighteenth century (Bognar 1985a, 1995b). As a consequence of changes in tectonic movements, river terraces are not continuous. Younger Pleistocene and Holocene terraces prevail. There are no older terraces due to subsidence. The exception is the upstream section of the Drava River (from

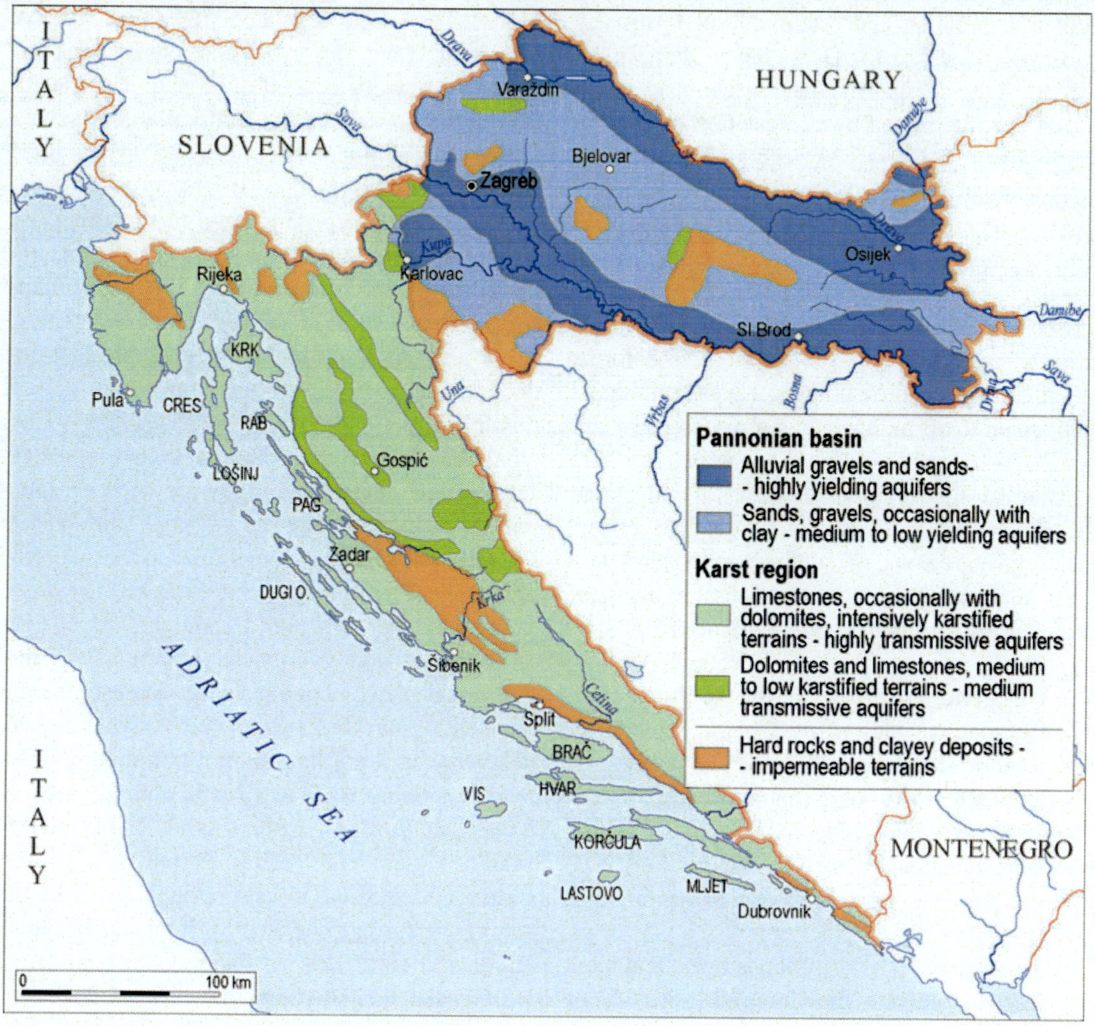

Fig. 1.3 Litho-hydrogeological map of Croatia (Modified after Majer and Prelogović 1993)

Koprivnica toward the Slovenian border), where four Pleistocene terraces and one old Holocene terrace have been identified. Furthermore, recent subsidence created the fluvial-marshy plains of Kopački Rit (Bognar 1985b).

Scattered mountains, hilly regions, pediments (glacis), and loess plateaus in the Pannonian Basin on the Tertiary and Quaternary clastic deposits are areas where intensive slope (sheet wash, solifluction, gullying, landslide, rockslide, and rockfall), pseudokarst, and fluvial processes are active (Bognar 2001).

Loess plateaus of eastern Croatia, with the domination of typical carbonate loess, are marked by the formation of *pseudokarst forms* (loess dolines, loess wells, loess chasms, loess pyramids, loess circuses, surducs, and steep loess sections along the banks of the Danube River) and corrosion-piping processes. Loess dolines, on the other hand, developed primarily due to corrosion, piping, and colluvial processes. Abrasion and fluvial erosion also contributed to the development of other landforms (Bognar et al. 2003).

In the Anthropocene, pastoral nomads cut down trees in the area that would come to be known as the Military Frontier in the seventeenth century, and consequently caused a revival of aeolian activity and the development of the youngest deflation and accumulation forms (by northwest winds) on the Drava River plain. In the second half of the nineteenth century, wind action was interrupted by afforestation on aeolian sands (*Đurđevački peski*) (Bognar 1992).

A smaller portion of Croatian karst is insular karst, scattered throughout the Pannonian Basin (Medvednica, Ravna Gora, Papuk), developed in Mesozoic and Neogene carbonates. It is characterised by a smaller depth of karstification (mostly up to 100 m). The main surface forms are small and shallow dolines, caves, and shafts (mostly up to 50 m long and deep) (Bognar et al. 2012).

Due to the regular flooding of the large floodplain area in Pannonian Croatia, river terraces, hilly regions of southern exposure, and loess plateaus have always been the areas of preference for settlement and agriculture.

1.2.1.2 The Dinarides (Dinaric System)

The Dinarides (Dinaric Alps, Dinaric Mountains, and Dinaric Mountain System) are narrowest within the territory of Croatia, mainly in the Gorski Kotar Region (Croatian Corridor), which has been an area of great transportation, economic, and strategic importance throughout history.

Carbonate rocks, mostly limestones, and karst landscapes dominate the Dinarides (40.6% of Croatian territory) and control drainage. Coastal mountain chains have the highest precipitation, but most water disappears underground because of fractured carbonate complexes.

Dolines, as one of the most typical and most widespread karst features, are particularly numerous in the Dinaric part of Croatia. In the geodatabase, 349,324 dolines have been recorded. The density is highest in Gorski Kotar (242 dolines per km^2). Dolines can appear individually, as is the case on the littoral slopes of the Dinarides and on different islands, but usually they form doline fields and long rows along contacts, on fractures, joints, and smaller faults. The areas with extreme density make some parts look like cockpit karst. Mountain tops are also characterised by megadolines, more than 100 m deep (e.g. at Rožanski Kukovi on Velebit Mountain[10]). The elevated relief of Velebit Mountain (and also Risnjak and Biokovo mountains) was glaciated during the last glacial maximum, so the pre-existing karst depressions transformed into glacial cirques and megadolines (Bognar and Faivre 2006; Bognar et al. 2012).

Poljes are typical landscape features in karst terrain with extensive flat bottoms used as arable land, with springs, ponors, estavelas, and sinking streams. The evolution of these large polygenetic depressions is determined by geological structure, karst, fluvial, and slope processes. Their dominant morphological characteristic is the elongated Dinaric axis (northwest to southeast), several times longer than they are wide. Despite regular inundations caused by differences between large inflow and limited ponor capacities, poljes are well-populated areas. They are surrounded by rocky hills and mountains covered with forests and grasslands, important for the traditional karst economy. The largest among them is Ličko Polje (465 km^2). The largest waterway is the Lika River (78 km long—the longest sinking river in Croatia and the second longest in Europe). After sinking 1,700 m into Markov Ponor by Velebit Mountain, it flows further underground toward springs and Adriatic submarine springs between St. Juraj and Jablanac (Bognar et al. 2012).

Because the karst environment of Mediterranean Dinaric Croatia is marked by rock permeability, scarcity of water, soil, and arable zones, as well as regular summer droughts, the population adopted pastoralism as the most adaptable economic activity and lifestyle throughout the centuries, in order to cope with the difficult environment. Excessive pastoralism under conditions of Mediterranean climate, thin soil cover, and especially steep slopes eventually led to the expressed deforestation (Fuerst-Bjeliš and Durbešić 2013). Especially valuable and very restricted landforms for tillage were (and still are) dolines and poljes.

[10]Simply referred to as Velebit by Croatians. It has multiple peaks but is considered to be a single, albeit rather extensive, mountain.

Erosional surface fragments are found at various levels in Croatia (southern Istria, northern Dalmatia, Zadvarje, and Lika), erosional surfaces in northern Velebit, and also in nearby countries (Bosnia and Herzegovina, Montenegro) from sea level to 1,000 m a.s.l. and even above 2,000 m a.s.l. They represent the traces of a once larger, complete, and older planation surface. During the neotectonic period, this surface was intensively disrupted and fragmented, and its preserved sections were transported to various hypsometric positions (Bognar 1996).

In the framework of the Dinarides, valleys of complex development prevail (alternating gorges and intermountain basins), due to neotectonic movements and heterogeneous carbonate (limestones and dolomites) deposits. Springs of the allogeneous streams are located in the mountainous hinterland, in impermeable rocks. As the streams make their way through the limestone–dolomite belt, the aforementioned springs help them maintain their flow despite losing water to porous rock. Canyons are characteristic for the areas of pure limestone (especially for the Čikola and Krka rivers, and somewhat for the Zrmanja and Cetina rivers). The river mouths at the Adriatic Sea are mostly of the *rias* type. *Deltas* are rare but an exception is the delta of the Neretva River (Bognar et al. 2012).

1.2.1.3 Coast and Sea-Level Change

The majority of the Croatian coastal area is formed in Cretaceous carbonates accumulated on the ancient Adriatic carbonate platform. Palaeogene limestones, calcareous breccias, and flysch are also widespread (Bognar et al. 2012).

The shape of the coast strongly depends on lithology. According to the prevalence of limestones and dolomites, erosional rocky coast predominates. Accumulation coasts are primarily linked with river valleys and torrents (Bognar et al. 2012). Several river mouths existed as large and deep sea bays until historical times. Rivers that pass only through karst areas, often in canyons, represent contemporary deep sea bays (e.g. the estuary of the Krka River) because karst areas do not contribute significantly to the river load (Janeković et al. 1995).

The human impact on the natural coastline is very strongly expressed today. Changes in beach equilibrium induced by human activities can be observed in numerous areas. Beaches that formed from the fan material of submerged torrent karst valleys can be anthropogenically affected, due to interventions along the watercourses that diminished the natural recharge of beach material (Faivre et al. 2011a; Faivre and Mićunović 2017). On the other hand, man-made constructions along the coast also supply a surplus of material that enlarges natural beaches (Rajčić et al. 2010).

The recent shape of the carbonate rocky coast is primarily a consequence of the Late Pleistocene–Holocene sea-level rise when the karstified relief was partially submerged (Fouache et al. 2000; Benac et al. 2004; Faivre et al. 2011b; Bognar et al. 2012). The relative sea-level changes were not uniform along the eastern Adriatic coast. During the last 2,000 years, different periods of relative sea-level change relate to periods of rapid climate change (Faivre et al. 2013, 2019). Evidence of the existence of temporary stillstands can be found in tidal notches (Faivre and Fouache 2003; Faivre and Butorac 2018) and on large algal rims (Faivre et al. 2013, 2019).

1.2.2 Climatic-Vegetation Areas

In terms of climate, there are expressed differences among different parts of Croatian territory, however, looking at climate and environmental criteria on the whole, we can define three essentially different wholes: peri-Pannonian/Pannonian, Dinaric, and Mediterranean (Fig. 1.4).

1.2.2.1 Peri-Pannonian/Pannonian Area

The *peri-Pannonian/Pannonian area* encompasses the Pannonian Basin, and has a moderately warm, rainy climate without dry periods (type Cfb according to Köppen). Air temperature in mid-July averages 20 to 22 °C, and from 0 to 3 °C in January. The amount of precipitation ranges from 800 to 900 mm (up to 1200 mm in the far western part of the region), and is evenly distributed throughout the year with the

Fig. 1.4 Climatic regions of Croatia (After Filipčić 1998)

minimum during the colder months. The greatest amount of precipitation comes in spring (May), and the second maximum comes in late summer (July or August).

Further to the east, the yearly amount of precipitation is lower and ranges from 600 to 700 mm, and is marked by a single maximum in the warm part of the year. One specific aspect of the peri-Pannonian/Pannonian area is the fluctuating amount of precipitation, due to which even a mild reduction can result in a drought. There is also a marked difference in the range of temperatures between the eastern and western parts of the area, which is a consequence of the expressed continental climate in the east. Due to this, the eastern part has higher air temperature values in the summer, marked by the occurrence of hot days, and low temperatures in the colder part of the year, marked by the frequent occurrence of cold days (Rogić 1990; Seletković and Katušin 1992; Šegota and Filipčić 1996).

Climate characteristics are linked to the development of different types of soil. Chernozems developed on a loose substrate have the greatest significance, primarily in the far eastern part of the area. This area boasts soil that is rich with organic substances and very well-developed humus. The production of a large quantity of organic substances in the upper levels of the soil is linked with the development of lush grassy vegetation in the warm and humid spring. Summer heatwaves, and a short autumn, do not allow for the strong development of microbiological processes and mineralisation of organic substances. Conversely, exceptionally low winter temperatures do not allow for the production or dissolution of organic substances. One characteristic of Chernozems, apart from the fact that the production of organic substances in the soil is greater than the dissolution thereof, is its sponginess, which is important for soil aeration and porousness. Chernozems are also developed in flood plains in the form of humus-rich Humic Gleysols. Because the western part of the area has a greater amount of precipitation, water leeches organic substances from the upper layers of the soil and degrades the humus, creating so-called degraded Chernozems. This process is more expressed in hilly terrain higher up on the relief, due to the fact that there are different types of eluvial and illuvial soils—even Podzols—which are typical for the Dinaric area (Rogić 1990).

Vegetation in the peri-Pannonian/Pannonian area has significantly changed throughout history due to human impact, i.e. agricultural use of the most valuable arable soil. The original natural vegetation of the western part of the area consisted of sessile oak (*Quercus petraea*) and European hornbeam (*Carpinus betulus*) forests. These forests developed vibrantly in an area of moderate humidity, and to this day they are still totally preserved in hard-to-reach and thinly settled hilly areas (Fig. 1.5). In contrast, on steep, south-exposed limestone slopes we find forests of downy oak (*Quercus pubescens*) and Oriental hornbeam (*Carpinus orientalis*). The high hilly areas of the peri-Pannonian region are covered with European beech (*Fagus sylvatica*) which occurs above the belt of oak forests.

In the highest areas, especially on soils formed on magmatic rocks, we find coniferous forests[11] (Rogić 1990). On the highly fertile Chernozems of the far eastern part of the area, east of the Slavonian mountains Bilogora, Papuk, Krndija, and Dilj, forest vegetation gives way to steppe. This transitional forest-steppe area has very little original plant cover, because it has largely been suppressed by human agricultural activity throughout history. River flood plains and swampy areas are characterised by hydrophilic forest and shrub vegetation, primarily European oak (*Quercus robur*), narrow-leafed ash (*Fraxinus angustifolia*), cottonwood (*Populus*), willow (*Salix*), and alder (*Alnus*) (Rogić 1990).

1.2.2.2 Dinaric Area

The *Dinaric area* encompasses the central mountainous area between the mountain ranges of Risnjak, Velika Kapela, Mala Kapela, Velebit, Plješevica, and Dinara. The Dinaric area also includes the area of mountainous Croatia, i.e. the regions of Lika and Gorski Kotar. This is also an area with a continental climate of type Cfb, similar to the Pannonian Basin. Even so, there are differences in climatic characteristics between mountainous Croatia and the Pannonian Basin. Namely, the climate of the Dinaric area also has no dry period, with a yearly amount of precipitation from 1,400 to 1,600 mm. Precipitation is distributed evenly throughout the year, with the minimum amount falling in the hottest months. On the other hand, the maximum amount of precipitation happens in late autumn, while the secondary maximum happens at the start of the warm part of the year. The average air temperature in July, the hottest month, ranges from 20 to 22 °C, and from 0 to 3 °C in January, the coldest.

It is important to state, however, that the area of the mountains themselves, which delimits the Dinaric area within Croatian borders, is characterised by a different type of climate. Namely, a boreal or snowy-forest climate (of the types described under section D, according to Köppen) that is found above 1,200 m a.s.l. Snowy and cold winters are characteristic, with average temperatures lower than -3 °C in the coldest months. Summers are brisk and short with average temperatures in the warmest months of

[11] This is a pine forest, while on the highest parts of Papuk, Medvednica, and Ivančica mountains there are continental beech forests.

Fig. 1.5 Forest vegetation communities (Modified after Bertović 1993)

above 10 °C (mostly between 15 and 20 °C). The amount of precipitation is significantly higher (over 1,800 mm) and is distributed throughout the entire year with the maximum in the colder months. The share of snow in total precipitation is significant, and snow blankets the soil for much of the year. The large amount of precipitation is a consequence of air currents, primarily warm ones that bring large amounts of rain and snow in the cold half of the year. Precipitation in the warm half of the year, conversely, comes about due to air currents from the Mediterranean and the Pannonian Basin, when the air cools and causes condensation, cloud development, and precipitation. An additional specific climatic characteristic of the Dinaric area is the occurrence of temperature inversion in both the warm and cold parts of the year, particularly during calm anticyclonic weather (Rogić 1990; Seletković and Katušin 1992; Šegota and Filipčić 1996).

Luvisols, of which Humic Podzols are most abundant, have developed on the primarily carbonate bedrock of the Dinaric area amid conditions of large quantities of water brought by precipitation and melting snow. Water washes away soluble minerals of the upper layer of the soil, after which they are deposited into deep soil layers. Accordingly, the soil is less fertile and lacking in nutrients, a situation that can be remedied by artificial means (fertilisers). On limestone mountains, we also find layers of terra rossa, upon which Podzols have also developed.

In contrast, in the highest mountainous areas there is a special kind of black soil which contains a larger quantity of humus, called Calcomelanosol. Its existence is linked to climatic conditions. Namely, a specific characteristic of the highest Dinaric mountainous areas is an expressed dryness of the surface layer of the soil during the warm part of the year due to wind. This slows microbiological processes and mineralisation of organic substances during the warm period. The cold of winter, on the other hand, prevents the production and dissolution of organic substances. Humus rich in mountain Calcomelanosols offers exceptional conditions for the development of grassy vegetation, in the form of forest glades and meadows. Therefore, in the highest areas, forest vegetation is often suppressed to create mountain pastures (Rogić 1990).

The high humidity and low air temperature best favour the spread of beech forests. They mix with oak forests in lower areas, and with fir forests (*Albies alba*) in higher areas. In the concave forms of the relief, primarily in large dolines, in which the air temperatures are even lower and the humidity still higher, there are Norway spruce forests (*Picea abies*). In the highest areas, which are also the coldest, the forest gives way to deformed and stunted woods.

1.2.2.3 Mediterranean Area

The *Mediterranean climatic–ecological area* encompasses the area of Mediterranean climate, as well as transitive sub-Mediterranean climate. The Mediterranean climate area (Csa according to Köppen) includes the narrow coastal belt of Dalmatia and the Dalmatian islands, and only goes deeper inland in the area of the Neretva Delta. This is the basic type of climate for the Mediterranean coast, characterised by the transition of a dry and hot half of the year, for which very hot days are common, to a cold and humid half of the year. Summer heatwaves are a consequence of expressed daily warming of an area that is relatively low in terms of relief.[12] The distribution of precipitation shows a concentration during winter (December and January), such that the most rainy winter months have three times the amount of precipitation than the driest summer months, which record less than 40 mm on average. The sub-Mediterranean belt includes Istria, Kvarner Bay, and the interior of Dalmatia (with the exception of the Neretva Valley).

This is a moderately warm, humid climate with hot summers. Average July air temperatures are typically ≥ 22 °C. The rainy season also overlaps with the cold part of the year, with maximums in spring (from April to June) and autumn (October and November). Yearly precipitation in the Mediterranean area ranges from 500 to 1,200 mm, mostly in the form of rain as a consequence of cyclonic currents in the cold part of the year. Snowfalls are rare and are linked to infrequent incursions of cold, continental air.

Karst hydrology in Mediterranean Croatia, i.e. the subterranean sinking of the majority of surface water, as well as powerful evaporation, causes an expressed lack of water during the warm part of the year. In summer months, anticyclonic weather prevails, while the cold part of the year is characterised by changing winds: the dry and cold bora (north) and the warm sirocco (south)[13] (Rogić 1990; Seletković and Katušin 1992; Šegota and Filipčić 1996).

The Mediterranean area is characterised by coloured soil types, such as terra rossa (Rhodic Cambisols) and Calcic Cambisols, and a significant area is simply covered in rocks. The soil here is less fertile due to the lower quantity of humus in its upper layers. The growth of plant cover and production of organic substances are hindered by high summer temperatures and low levels of moisture in the soil. In contrast to this, due to mild air temperatures and high levels of precipitation in the cold part of the year, the

[12]Although the Adriatic sea has a cooling influence during the summer months, summers are characterised by high air temperatures, showing the strength of this warming effect.

[13]The Croatian name for the bora wind is *bura*, and the scirocco wind is called *jugo*.

small amount of organic substances found in the soil become dissolved. Terra rossa is a relic soil created by the dissolution of limestone, which came into existence in the Miocene, and persisted through the Late Pleistocene under expressed subtropical climate conditions. Iron and aluminium oxides give the soil its red colour. As a result of contemporary climatic conditions, primarily cold air temperatures and large amounts of moisture, terra rossa has completely transformed into humus-enriched cambisols (Rogić 1990).

In the Mediterranean area, vegetation is primarily xerophyte, acclimatised to summer drought, high temperatures, and high levels of solar radiation. Primary vegetation consists of holm oak (*Quercus ilex*) forests and other coniferous plants, such as various kinds of pine (*Pinus*) and cypress (*Cupressus*) trees. Today, these forests can be found only in individual, spatially limited localities. Namely, the original natural forest has been exposed to thousands of years of human influence, overgrazing, farming, and logging.

Degradation of the original natural forest forms a characteristic degradation stage of Mediterranean vegetation called *maquis*.[14] Further degradation (grazing, logging, and fires) leads to a less-dense *garrigue* stage, which consists of the toughest forms of xerophyte bushes and grasses, and eventually leads to bare rocky karst. As a result of long-term human influence in the ecologically sensitive Mediterranean setting, degraded forests—*maquis* and bare rocky karst—make up the majority of the contemporary area.

Vegetation in sub-Mediterranean climate areas, however, is characterised overwhelmingly by deciduous forests dominated by pubescent oak (*Quercus pubescens*) and Oriental hornbeam (*Carpinus orientalis*). In the transition to the Dinarides, there are coastal beech forests (*Fagus sylvatica*) and degraded vegetation in which stubby forest plants like mugo pine (*Pinus mugo*) can be found (Rogić 1990).

[14]The term differs regionally. The Italian variant is *macchia*; in French it is *maquis*; in Spanish *matorral*; in Greek *phrygana*; in Portuguese *mato,* etc. It is a shrubland biome in the Mediterranean, typically consisting of dense growths of evergreen shrubs. Further degradation leads to the less-dense vegetation called *garrigue*.

References

Benac Č, Juračić M, Bakran-Petricioli T (2004) Submerged tidal notches in the Rijeka Bay NE Adriatic Sea: indicators of relative sea-level change and of recent tectonic movement. MarGeol 212(1–4):21–33

Bertović S (1993) Natural vegetation. In: Klemenčić M (ed) A concise atlas of the Republic of Croatia and the Republic of Bosnia and Herzegovina. The Miroslav Krleža Lexicographical Institute, Zagreb

Boban Lj (1995) Hrvatske granice od 1918. do 1993. godine. Školska knjiga – HAZU, Zagreb

Bognar A (1985a) Basic geomorphological problems of the Drava river plain in SR Croatia. Geogr Pap 6:99–105

Bognar A (1985b) Basic geomorphologic characteristic of the Kopač swamp. Geogr Glas 46:5–20

Bognar A (1992) Pediments of the South Velebit mountain range. Geogr Glas 54:19–32

Bognar A (1995a) Fizičko-geografske pretpostavke regionalnog razvoja Hrvatske. In: Pepeonik Z (ed) Zbornik radova I: hrvatskog geografskog kongresa, Hrvatsko geografsko društvo, Zagreb, pp 51–65

Bognar A (1995b) Regulations and his influences on the geomorphologic development of the Drava and Danube rivers in Croatia. In: Croatian conference of the waters, Dubrovnik

Bognar A (1996) Croatia—the land and natural features. GeoJournal 38(4):407–416

Bognar A (2001) Geomorfološka regionalizacija Hrvatske. Acta Geogr Croat 34(1999):7–29

Bognar A, Faivre S (2006) Geomorphological traces of the younger pleistocene glaciation in the central part of the Velebit Mt. Hrvat Geogr Glas 68(2):19–30

Bognar A, Faivre S, Buzjak N, Pahernik M, Bočić N (2012) Recent landform evolution in the dinaric and pannonian regions of croatia. In: Lóczy D, Stankoviansky M, Kotarba A (eds) Recent landform evolution the carpatho-balkan-dinaric region. Springer, Dordrecht, pp 313–344

Bognar A, Schweitzer F, Szőőr G (eds) (2003) Susak—environmental reconstruction of a loess Island in the Adriatic. Geographical Research Institute—Hungarian Academy of Sciences, Budapest

Bösendorfer J (2010) Nešto malo o našoj Baranji. Zavod za baranjsku povjesnicu, Beli Manastir

Butlin RA (1993) Historical geography: through the gates of space and time. Edward Arnold, London

Crljenko I (2014) Kulturnogeografska obilježja dinarskog krša Hrvatske. Acta Geogr Croat 39:15–44

Faivre S, Bakran-Petricioli T, Barešić J, Horvatić D, Macario K (2019) Relative sea-level change and climate change in the Northeastern Adriatic during last

1.5 ka (Istria, Croatia). Quaternary Science Reviews 222. https://doi.org/10.1016/j.quascirev.2019.105909

Faivre S, Bakran-Petricioli T, Horvatinčić N, Sironić A (2013) Distinct phases of relative sea level changes in the central Adriatic during the last 1500 years—influence of climatic variations? Palaeogeogr Palaeoclim Palaeoecol 369:163–174

Faivre S, Butorac V (2018) Recently submerged tidal notches in the wider Makarska area (Central Adriatic, Croatia). Quat Int, Quat Croat 494:225–235. https://doi.org/10.1016/j.quaint.2017.07.020

Faivre S, Fouache E (2003) Some tectonic influences on the Croatian shoreline evolution in the last 2000 years. Z Für Geomorphol 47(4):521–537

Faivre S, Fouache E, Ghilardi M, Antonioli F, Furlani S, Kovačić V (2011a) Relative sea-level change in Istria (Croatia) during the last millennia. Quat Int 232:132–143

Faivre S, Pahernik M, Maradin M (2011b) The Gully of Potovošća on the Island of Krk—the effects of a short-term event. Geol Croat 64(1):64–76

Faivre S, Mićunović M (2017) Reconstructing recent beach morphological changes by repeat photography—example of Zogon beach on the island of Hvar (Central Adriatic). Geoadria 22(2):165–192

First release, Year LIV, 14 Sept 2017, Croatian Bureau of Statistics.

Fouache E, Faivre S, Dufaure J-J, Kovačić V, Tassaux S (2000) New observation on the evolution of the Croatian shoreline between Poreč and Zadar over the past 2000 years. Z Geomorph NF Suppl-Bd 122:33–46

Fuerst-Bjeliš B, Durbešić A (2013) Littoralization and behind: environmental change in Mediterranean Croatia. In: Pina H, Martins F, Ferreira C (eds) The overarching issues of the European space/strategies for spatial (Re)planning based on innovation, sustainability and change. Fundação Universidade do Porto, Porto, pp 136–147

Hughes JD (2011) Što je povijest okoliša? Disput, Zagreb

Janeković M, Juračić M, Sondi I (1995) Sedimentacijske osobitosti rijeke Mirne (Istra, Hrvatska). In: Vlahović I, Velić I, Šarica M (eds) 1. hrvatski geološki kongres—zbornik radova 1, Institut za geološka istraživanja, Zagreb, pp 225–227

Klemenčić M (1991) Novija historijsko-geografska osnovica jugoslavenskih izvanjskih i unutarnjih granica s posebnim osvrtom na hrvatske granice. In: Crkvenčić I (ed) Političko-geografska i demografska pitanja Hrvatske. Savez geografskih društava Hrvatske, Zagreb, pp 317–334

Majer D, Prelogović E (1993) Hydrogeology and seismotectonics. In: Klemenčić M (ed) A concise atlas of the Republic of Croatia and the Republic of Bosnia and Herzegovina. The Miroslav Krleža Lexicographical Institute, Zagreb

Rajčić ST, Faivre S, Buzjak N (2010) Preoblikovanje žala na području Medića i Mimica od kraja šezdesetih godina 20: stoljeća do danas. Hrvat Geogr Glas 72 (2):27–48

Rogić V (1971a) Retrospektivno i suvremeno značenje prometne funkcije ličko-goranske regije. In: Roglić J, Božičević J (eds) Naučno savjetovanje Prometna valorizacija Hrvatske, Zagreb

Rogić V (1971b) La Croatie dinarique. Bull De La Fac Des Lett De Strasbg 49:210–231

Rogić V (1972) Espace vecu et societes en Croatie. Bull De La Fac Des Lett De Strasbg 50:211–240

Rogić V (1975) Regionalno-geografska izdvojenost i geografske osobine riječkog prostora. In: Cvitanović A (ed) Geografija SR Hrvatske—knjiga V: Sjeverno hrvatsko primorje. Školska knjiga, Zagreb, pp 79–97

Rogić V (1976a) Socio-geografski aspekt dinarskog krša, dinarskog kulturnog areala i dinarskog brdsko-planinskog prostora. Geogr Glas 38(1):253–269

Rogić V (1976b) Socijalno-geografski aspekt primorskog krša, In: Roglić J (ed), Simpozij Ekološko valoriziranje primorskog krša, Split—Zagreb

Rogić V (1982) Historijsko-geografska osnova socijalno-kulturne diferencijacije vojnokrajiškog prostora. Geogr Glas 44(1):23–38

Rogić V (1990) Regionalna geografija Jugoslavije knjiga 1—prirodna osnova i historijska geografija. Školska knjiga, Zagreb

Roglić J (1950) O geografskom položaju i ekonomskom razvoju Jugoslavije. Geogr Glas 11–12:11–26

Schenk W (2011) Historische geographie. Wissenschafliche Buchgesellschaft, Darmstadt

Seletković Z, Katušin Z (1992) Klima hrvatske. In: Rauš Đ (ed) Šume u Hrvatskoj. Šumarski fakultet Sveučilišta u Zagrebu—Hrvatske šume, Zagreb, pp 13–18

Statistical Yearbook (2017). Croatian Bureau of Statistics, Zagreb

Šegota T, Filipčić A (1996) Klimatologija za geografe. Školska knjiga, Zagreb

Williams M (1994) The relations of environmental history and historical geography. J Hist Geogr 20(1):3–21

2 Prehistory—The Early Interactions Between Humans and Their Environment

Abstract

The first human–environmental interactions prior to the Holocene date back to the Paleolithic. There are 50 localities throughout Croatian territory in which we find traces of Paleolithic human life. The rich assortment of animals enabled the development of hunting as a primary activity, while gathering was highly seasonal. Neolithic humans had an unprecedented influence on the natural landscape in terms of introducing agriculture and consequential sedentarisation. The concentration of Neolithic society within two areas—the Danubian area (with a strong agricultural component) and the Adriatic–Mediterranean area (boasting a strong herding component)—is one of the era's main characteristics. The first instances of deforestation in the area of Croatia started during the Neolithic period. Forest communities (primarily oak) were suppressed for land cultivation—based on burning forests in order to create arable land. More than 3,000 years of a pastoral economy based on sheep and goat herding in the Adriatic–Mediterranean area had a very destructive effect on forest cover, and eventually resulted in the degradation of (peri-) Mediterranean forests into *maquis* or bare rock. With the transition to the Aeneolithic, environmental priorities changed and the Dinarides became a favoured area. Numerous remains of material culture from this era bear evidence to this: first and foremost the numerous fortifications (hillforts) that Indo-European settlers raised on difficult to reach parts of the relief, which were ideal for the defense and control of the surroundings. Consequently, in the metal ages, the Dinaric area was eventually affected by the degradation of forests as well.

Keywords

Croatia · Prehistory · Neolithic culture · Danubian area · Adriatic–Mediterranean area · Deforestation · Hillfort · Vučedol · Illirian tribes · Illyro-Celtic tribes

B. Fuerst-Bjeliš and N. Glamuzina, *The Historical Geography of Croatia*, Historical Geography and Geosciences, https://doi.org/10.1007/978-3-030-68433-4_2

Monkodonja: prehistoric proto-urban settlement and hill-fort in detail, western Istria

Hillforts are prehistoric settlements and fortifications of Indo-European tribes, built in high places for strategic purposes. Monkodonja is a Bronze Age settlement (1,000 inhabitants) and an important strategic point in communication between Central Europe and the Mediterranean/Aegean area. The archaeological site is the award winner of Europa Nostra for the exemplary excavation, study, and presentation (Photograph: Ivan Zupanc. Reproduced with permission)

Significant interactions between humans and their natural environment began during the prehistoric era. The traces of these interactions are superimposed upon one another throughout history like a palimpsest and form various types of cultural landscapes. Although there are no written documents from this era, thanks to scientific research and knowledge stemming from numerous sister disciplines like archaeology, paleoclimatology, stratigraphic geology, anthropology, geobotany, and zoography, it is possible to reconstruct the relationship between humans and the environment, and thus the shaping of the first cultural landscapes. Significant, and perhaps key, findings can also be found by applying comparative methods to the analysis of societies still living at the cultural level of the Stone Age. Namely, there are numerous results from research of Paleolithic and Neolithic societies from distant parts of Africa, and the tropical rainforests of Australia, Southeast Asia, and Central and South America. Cultural studies of communities like the Bushmen of South Africa, transhumant herders from northern and eastern Africa, hunter/gatherers from sub-polar areas (Eskimos, Saami, and peoples of northern Siberia), or Mongolian nomads are especially important in this regard (Jäger 1969). In relation to the influence of human activity on the environment, the prehistoric era can be divided into two relevant periods: the Stone Age and the Metal Ages (Bronze and Iron).

2.1 Paleolithic—Human Interactions with the Environment Prior to the Holocene

The Stone Age is conventionally divided into the Paleolithic and Neolithic, while the transitory period between the two is called the Mesolithic. The Paleolithic encompasses the greater part of prehistory, lasting from 2.6 million to 10,000 years ago. The beginning of the period overlaps with the first known usage of stone tools by early hominids, and the end with the close of the last ice age (Pleistocene). Migrations of neanderthals and probably younger communities of *Homo sapiens*[1] into what is today the European continent, regardless of their small numbers and limited abilities, marked the first human influence on periglacial Pleistocene forest and steppe environments. In this period, the existence of human society was based on gathering and hunting, which did not have a lasting impact on the natural landscape as human society was scattered into small hunter/gatherer groups.

According to scientific assessment, groups of around 50 people needed 1,500 km^2 to ensure survival. The first human communities usually used natural objects like caves as places to sleep and hide from bad weather. In areas where such natural protection was not available, as is the case in most lowland areas, humans lived in camps under the open sky (Leroi-Gourhan 1983). During the Pleistocene in contemporary Croatia, there was a dichotomy of environmental characteristics between coastal and continental areas. The coastal belt, along with the mountainous area and its hinterland, was characterised by communities of Mediterranean xerophyte shrubs and mountain forests, as well as various mammalian communities: hamsters (*Mesocricetus newtoni*), rodents (*Dolomys*), leopards (*Panthera pardus*), and lynx (*Lynx lynx*).

On the continental side, the Pannonian Basin belonged to the Pontic–Caspian Steppe which was dominated by animal species of the families of reindeer (*Rangifer tarandus*), mammoths (*Mammuthus*), rhinoceros (*Coelodonta*), antelope (*Saiga*), musk oxen (*Ovibos*), steppe bison (*Bison priscus*), horses (*Equus*), cave lions (*Panthera spelaea*), rodents (*Dicrostonyx, Allactaga*), lemmings (*Lemmus*), marmots (*Marmota*), steppe polecats (*Mustela eversmanii*), and Irish elk (*Megaloceros giganteus*) (Baryshnikov and Markova 1992). This rich assortment of animals enabled the development of hunting as a primary human activity. Gathering, on the other hand, was

[1]Of the twelve species of the genus *Homo*, which belong to the greater hominid family, eleven of them died out over the course of evolution, and only *Homo sapiens* (modern humans) endured.

highly seasonal and oriented toward shellfish, fruit, berries, tubers, mushrooms, chestnuts, oak and beech acorns, plant sprouts, roots, bulbs, grains, tree cores (especially pine and birch), and leaves (most often birch).

It should be noted how plants most likely had the role of enhancing basic nutrition, due to both lesser caloric value (especially in the colder Pleistocene climate) and because, unlike hunting, gathering was not practical throughout the entire year (Leroi-Gourhan 1983). There are 50 localities registered in Croatian territory in which we find traces of Paleolithic human life. The sites are spatially concentrated in two continental regions: Hrvatsko Zagorje and Požega Valley, and in three regions in the Mediterranean: Istria, Kvarner Bay, and north and central Dalmatia between the Cetina and Zrmanja rivers (Regan 2003).

2.2 Human Relations with the Environment in the Neolithic

The Pleistocene Ice Age[2] ended 10,000 years ago when the climate began to warm. Once the warming was underway, much of the ice covering parts of Eurasia and North America began to melt. The great ice sheet that had covered Europe during the Pleistocene receded to the German Mittelgebirge and the northern edge of the Carpathian Mountains, while south of that line, large swaths of ice only remained in isolated mountain ranges (the Alps, Massif Central, and Pyrenees) (Fig. 2.1). In contemporary Croatia, traces of glaciation can be found only on individual high peaks of Velebit Mountain, where smaller glaciers were formed (Bognar and Faivre 2006).

Due to the melting ice, the world sea level rose 96 meters in relation to the average sea level during the Pleistocene (Šegota 1982). Therefore, the world's contemporary coastline—Croatia's beautiful coast included—was shaped during the transition from Pleistocene to Holocene. As sea levels rose, low-lying terrain in the contemporary northern Adriatic flooded, as well as the valleys between mountainous ridges, creating the Adriatic islands (Šegota and Filipčić 1996). Simultaneously, the shaping of hydrographic networks and the contours of the relief began, whereby exogenic processes (fluvial, slope, karst, etc.) had a primary role. Forest vegetation began to spread and shaped the primary (natural) landscape during the Holocene. Beech forests were dominant in the mountainous interior, oak forests of the Mediterranean and peri-Mediterranean type in the littoral, and oak forests of the continental type (with a mixture of grass communities, i.e. steppe) in the Pannonian Basin.

The large number of relic plant species, which are characteristic of continental forests of sessile oak and European hornbeam in the Pannonian Basin, point to the special position that the territory had in that era. This area was a hot spot for the development of Holocene forest vegetation in Southeast Europe. In concurrence with climate change, vertical elevation shifts of the main forest-vegetation zones occurred. During the cold and arid Preboreal period (eighth and seventh millenniums BC), the elevation of vegetation zones was several hundred meters lower in altitude than today. A mild and dry boreal climate began at the end of 7000 BC. Temperatures became milder and more humid throughout the era, called the Holocene climatic optimum (between 6000 and 3000 BC), when vast forests covered contemporary Croatia. Mediterranean forest cover spread deeper into the continental interior with the warming temperatures.

In the Pannonian Basin, beech and oak forests were mostly dispersed in hilly and low-lying drained areas. Up to 500 m a.s.l., there was a transition zone between drained areas of steppe-forest vegetation and humid areas of rich forest vegetation. The vibrant development of vegetation almost entirely erased the humble traces of Paleolithic humans' interactions with their natural environment. Accordingly, during the Holocene climatic optimum, the basis for today's natural landscape came into being (around 5000 BC), setting the stage for later millennia of gradual transformation. Throughout history, humans have

[2]The end of the last Würm glaciation (Würm III) is taken as the end of the Pleistocene (the last ice age) and start the Holocene (the current geological era).

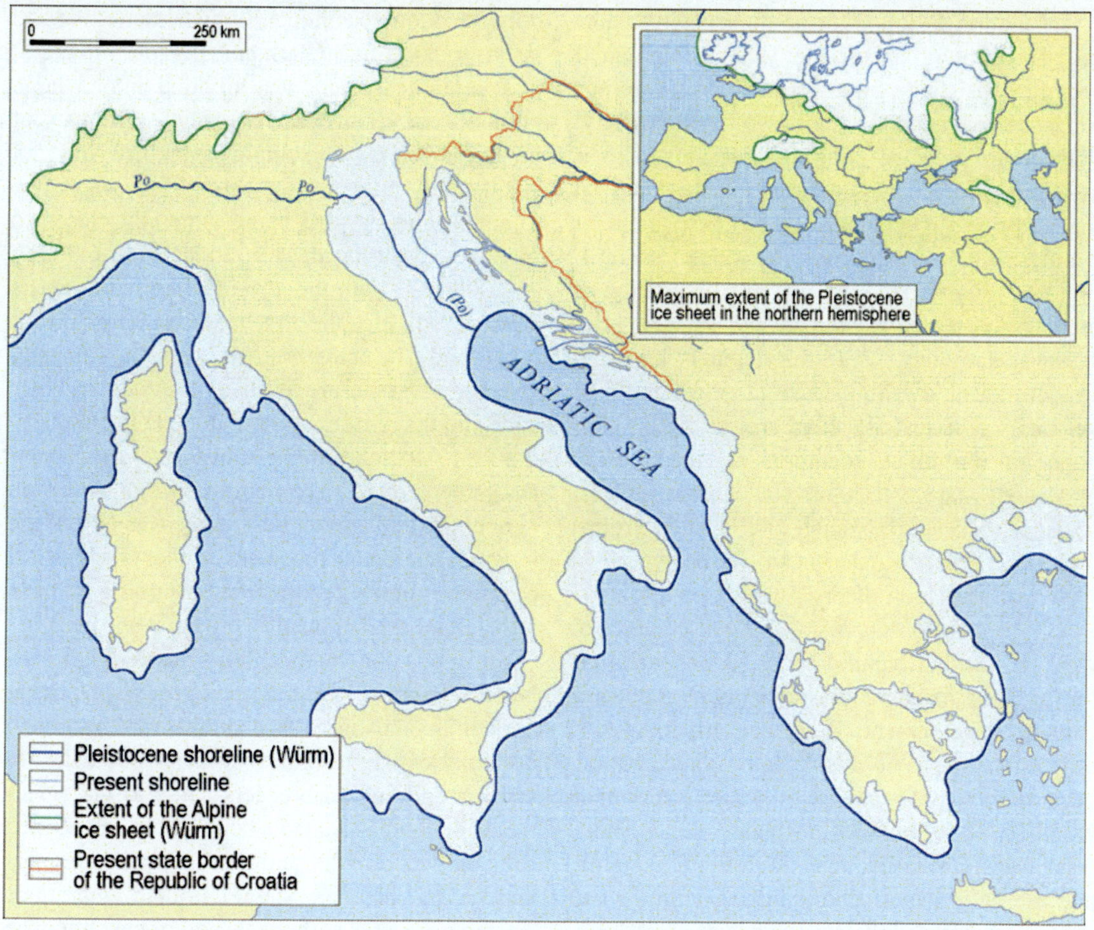

Fig. 2.1 Presumed Pleistocene shoreline in the Adriatic Basin and the maximum extent of the ice sheet (Würm). *Source* Dugački and Regan (2018)

modified the successive development of vegetation, mainly toward lower (regressive) development stages (Rogić 1990). The extinction of Pleistocene animal species and migration of other individual species toward northern areas with cold climates spurred climatic and vegetation changes. Additionally, it is very likely that humans significantly contributed to the disappearance of individual animal species on the European continent. This has also been confirmed by the fact that individual species from the *Felidae* family (cats), such as tigers and panthers, succeeded in surviving in Central Asia, despite having a harsher climate in relation to much of Europe (Leroi-Gourhan 1983).

Just how successful Paleolithic hunter/gatherers were at adjusting to new living conditions remains an open question. There are two broad hypotheses that attempt to answer this: the first regards autochthony, i.e. the gradual and successful adaptation of humans to new living conditions, while the second is based on the crucial role of migrations by which new groups, who were already adapted to new conditions, arrived in a given area. Research of cultures which developed during the Mesolithic gives more weight to the former hypothesis (Rogić 1990). The question of autochthony and migration is still an active topic of scientific discussion, being that the transition from the hunter–gatherer phase to more advanced forms of agriculture represents one of the most important turning points in human history. Migration theories arise from scepticism regarding the possibility of cultural advancement of societies based on a hunter–gatherer economy.

Population factors indicate their lifestyle during the greater part of the year within small groups, consisting of a small number of families (or just one), under conditions of food scarcity. Such groups needed large areas to satisfy their need for food. This is confirmed by research on the lifestyle of contemporary hunter/gatherer cultures—such as Eskimos, who need six million km^2 to ensure the survival of a population of 50,000. Additionally, hunting and gathering are seasonal activities, while migration is hindered by permanent dwellings and large amounts of personal possessions. Furthermore, the facts regarding unfulfilled conditions for the division of labour (specialisation) of hunter/gatherer societies, divided into small and isolated groups with very little mutual contact, further speak in favour of migration theory.

Due to this, conservatism, as a consequence of weak or lacking technological advancement, is one of the main characteristics of hunter/gatherer cultural development. Also, the psychological dimension must be taken into consideration, because the day-to-day life of hunter/gatherers is based on tactics rather than strategy, i.e. on short-term decisions rather than long-term planning, and the very acts of domesticating animals and cultivating plants fall into the category of long-term planning and represent a broadening of perspective. The theory regarding the domestication of dogs, to help with hunting, is still in question; however, it has been confirmed that the process of domesticating dogs resulted in dramatically enhanced human hunting capability, making hunting-based economies more efficient. On the other side of the coin, coinciding with climatic and ecological changes at the end of the Pleistocene, technological advancement of human society in areas of temperate climate (Europe, Southeast Asia) began. Furthermore, the Pleistocene overkill[3] was linked to the development of new crop cultivation and animal domestication techniques, which started at the latest in 9,000 BC (Clark and Piggott 1965).

Paleolithic society was divided into a large number of smaller, separate groups of hunter/gatherers. Paleolithic man also did not significantly influence the natural landscape, due to small numbers and being generally closed to innovation. During the Neolithic (from 6,000 to 2–3,000 BC), with the advent of agriculture as a new way of life, the first major cultures developed and the environment was significantly changed. Agriculture enabled a sedentary lifestyle and the creation of new spatial elements, i.e. the first settlements. The transition from a hunter–gatherer type of economy to one based on agriculture and animal husbandry was one of the most revolutionary processes in the development of human society, and marked a turning point in humankind's relationship with its environment.

In opposition to Paleolithic hunter/gatherers, Neolithic society was based on groups practicing agriculture and animal husbandry, which came to dominate certain natural resources. Soil, that early farmers and animal breeders came to via agreements or conquests, had a key role. In cases when the edges of usable land were reached, farmers and animal breeders started limited migrations toward new areas with arable soil or pastures. The agriculture/animal husbandry type of economy, as opposed to hunter–gatherer, depended heavily on the development of new agricultural techniques due to intense use of the existing land or the creation of new agricultural areas. Due to this, economic advancement was linked to the application of different innovations, as well as with the development of social mechanisms intended to save time and energy essential for effective agricultural production (Alexander 1977).

Some scholars, when analysing Neolithic society, put social components first and emphasise how the transition to food production is possible via social impetus—the emphasis here is placed on stronger social structure, development of social mechanisms for keeping order, as well as the need to vanquish enemies (Orme 1977). The first agricultural cores were developed in the temperate climate belt of Asia, where, starting in 5,000 BC, people began growing wheat and

[3]The overkill hypothesis argues that humans were responsible for the Late Pleistocene extinction of megafauna in northern Eurasia and North and South America. Some scholars see a chronological and causal link between humans' rise to prominence and the disappearance of many species of large mammals.

barley in the central and southeastern parts of the continent. Later, starting in 3,000 BC, the development of agriculture and animal husbandry began in three belt areas in Africa: the Mediterranean coast, the eastern highlands, and the Sahel. From these core areas, agriculture and animal husbandry began to spread to other parts of Africa and Asia, as well as into Europe (Alexander 1977). The first Neolithic agricultural core areas to emerge in Europe were found in the eastern part of the continent, in the continual zone stretching from the Aegean Sea to the Caspian Sea, spreading from there to the west.

The transition to a stationary and sedentary lifestyle was closely linked to the formation of the earliest settlements. The expansion of the first forms of crop cultivation to the west was linked to the problem of creating and continually maintaining these early settlements. The development of permanent settlements depended on secure sources of food, which could ensure the survival of their inhabitants for generations. Accordingly, grains that exhausted the fertility of the soil quickly were a significant problem to be overcome. The solutions were migratory agriculture, or natural regeneration of the soil using mud or irrigation (Clark and Piggott 1965). Becoming sedentary represented a revolutionary change in human lifestyle, much like the closely linked first agricultural revolution. With further advancement of shifting agriculture, the first permanent settlements stabilised, and after them came the first cities in Southwest Asia.[4]

2.2.1 The Development and Definition of the First Settlement Zones in the Neolithic

Neolithic humans had an unprecedented influence on the natural landscape. Within contemporary Croatia, we can identify two different areas (Fig. 2.2) in regard to the natural environment, lifestyle, and process of landscape transformation:

1. *Danubian* (continental)—in the eastern part of contemporary Croatia, with a strong agricultural component in relation to herding and hunter-gathering. The remains of many Neolithic and transitional Aeneolithic[5] cultures testify to this development: Starčevo (sites in Vučedol, Đakovo region, Virovitica in Podravina region, and other parts of Slavonia), Vučedol, Korenovo (the main site is situated in the surroundings of Virovitica and Požega Valley), Sopot (Vučedol, Lovas, Vukovar, Dalj, Sarvaš, and Osijek), and Vinča (Bapska, Vukovar, and Vinkovci).
2. *Adriatic–Mediterranean*—boasting a strong herding component in regard to agriculture and hunter-gathering. This development took place within the framework of cultures which encompassed the entire coastal belt: so-called Impresso pottery (main localities located in Istria and northern Dalmatia), Danilo (the core of this culture was located in the hinterland of contemporary Šibenik in the area of the settlements Danilo and near Vela Špilja cave on Lošinj), and Hvar (Vela, Markova and Grapčeva caves on Hvar island, Vela Špilja cave on Korčula, and Vela Špilja cave on Lošinj). The ceramics originating from other parts of Europe (from Sicily, the southeastern edge of the Alps, within the Dinaric belt, and from the eastern Mediterranean) that have been found here clearly demonstrate the development of trade within the Adriatic–Mediterranean area, directly or indirectly (Novak 1944).

The first deforestations in the area of Croatia, as well as in other parts of Europe, started during the Neolithic. Due to the technically limited abilities of Neolithic society, these processes were best achieved in labile zones such as forests or steppes. Neolithic humans suppressed forest communities (primarily oak), which had developed on very fertile loess, with the application of

[4]According to archaeological findings, the first urban settlements were formed in the Jordan River valley ('Ain Ghazal, Beidha, Jericho) and in Asia Minor (Çatalhöyük).

[5]The Aeneolithic is transitional period between the Neolithic and the Bronze Age in which some copper was used.

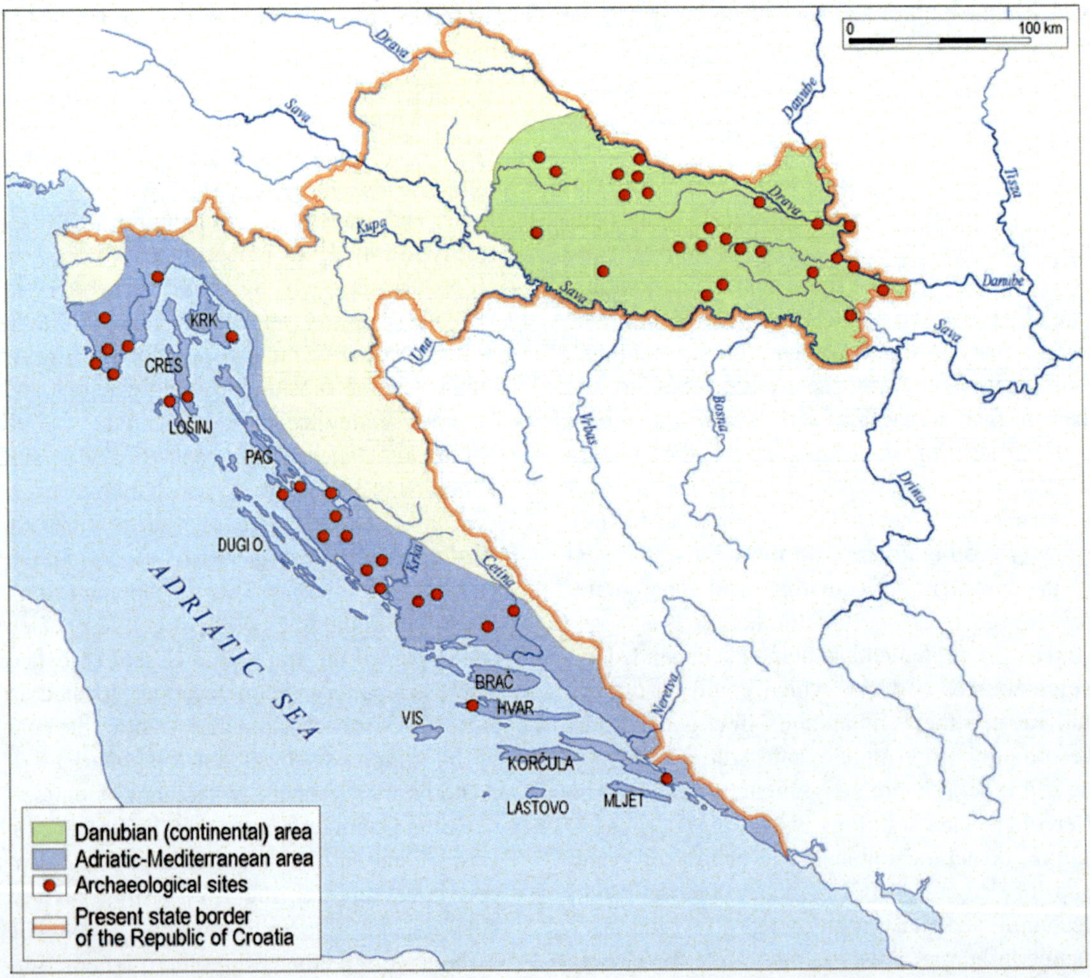

Fig. 2.2 Areas occupied by humans during the Neolithic. *Sources* Regan (2003), Dugački and Regan (2018). Modified after Glamuzina and Fuerst-Bjeliš (2015)

slash-and-burn agriculture. This primitive kind of land cultivation was based on burning forests in order to create arable land and increase the fertility of the soil with the resulting ash. Due to the fact that ash didn't add much to the fertility of the soil from year to year, it was necessary to continually burn new areas of forest to make new areas of fertile land.

Under conditions of labile bioclimatic–ecological balance, these sometimes-dominant forest communities did not succeed in regenerating their initial form. Apart from those first primitive forms of agriculture, the pastoral economy had an even greater effect on forest cover degradation. Namely, the first forms of the pastoral economy were based on extensive pasturing which further destroyed forest communities. The pastoral economy in the Adriatic–Mediterranean area, based on sheep and goat herding, had the most destructive effect on forest cover. This is the reason that forest vegetation became so degraded in the karst environment. More than 3,000 years of burning and destroying forests resulted in different forms of degradation of (peri-) Mediterranean forests into *maquis* or bare rock. The removal of vegetation also had a key role in the development of destructive processes which led to partial or complete degradation of the soil in the karst environment.

The Dinarides are located between the Adriatic–Mediterranean and Danubian areas and

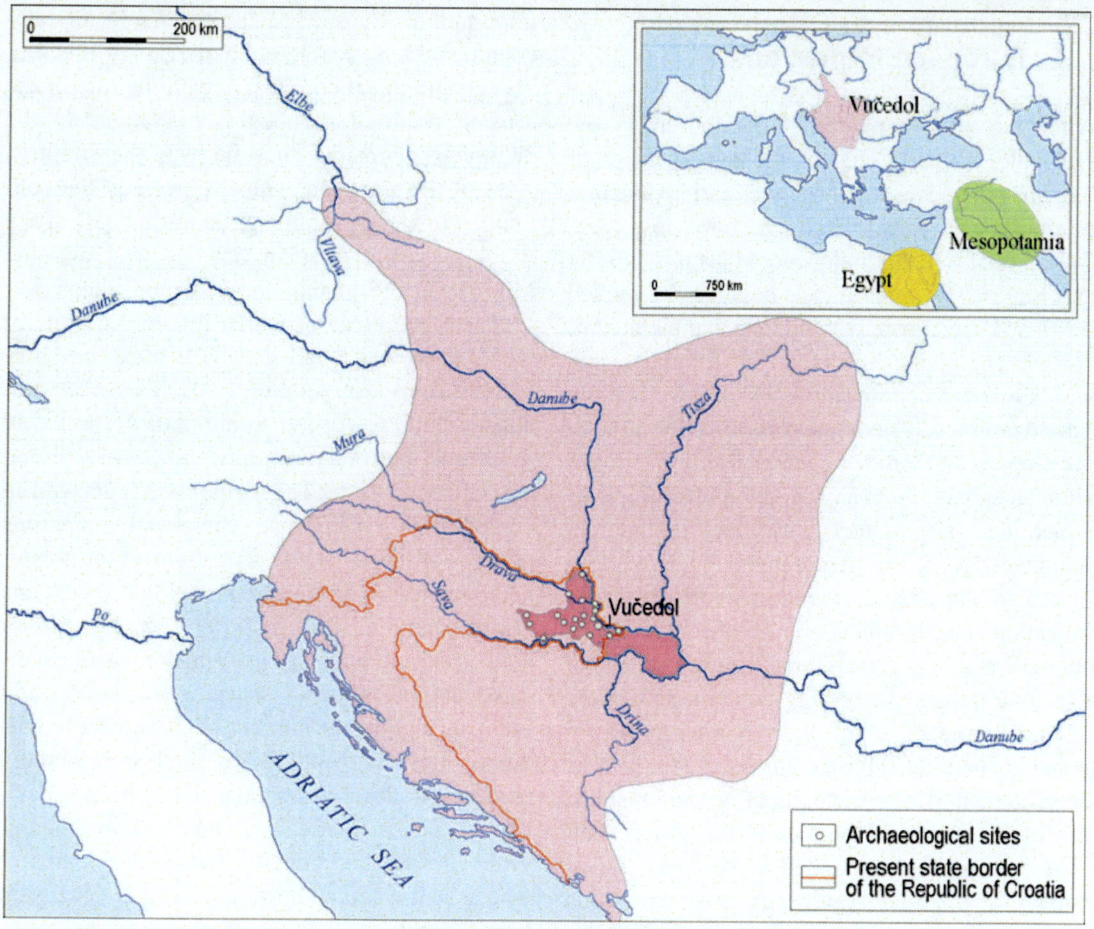

Fig. 2.3 The core and extent of the Aeneolithic Vučedol culture in Europe and concurrent ancient civilisations of the Old World. *Sources* Regan (2003), Dugački and Regan (2018). Modified after Glamuzina and Fuerst-Bjeliš (2015)

serve as a link between them. This is an area that was covered in dense beech forests and difficult to access and move through. As such it had a secondary role in the life of Neolithic humans, which is demonstrated by the scarce archaeological findings that indicate a sparse population. Therefore, the concentration of Neolithic society within the Danubian and Mediterranean areas is one of its main characteristics. General geographical dispersion, apart from thousands of years of influence on the natural landscape, is a characteristic of Neolithic society, which is shown by archaeological traces even on the distant open-sea islands (Palagruža and Svetac). Uninterrupted and long-lasting peaceful cultural development undoubtedly influenced the development of a sedentary, pacifist mentality which was most likely closely linked to a weaker will to fight and lesser mobility, and did not favour the development of geographically larger and socially more complex organisations (Fig. 2.3).

The fact that there were no significant traces of fortified settlements found until the late Neolithic, as well as little evidence of highly developed weapons, also supports this conclusion. Therefore, the notion that the eventual destruction of Neolithic society was closely linked with its cultural characteristics is justified (Rogić 1990).

2.3 Geographical Origins of Indo-European Migrations

After the first agricultural revolution in the Neolithic, the next step in the development of human society was the discovery and application of metalwork. Newer findings show how there were probably many independent centres in Asia and Europe, in which metals began to be melted down and used, from which metal and metalwork spread. The application of metal in hunting and war is linked to civilisations which developed in Southwest Asia. The transition from Neolithic to the Copper and Bronze ages probably happened simultaneously in many geographical areas—which had no contact with one another. In Southwest Asia, the beginning of metalwork is linked to the area of the Iranian Plateau. The invention and application of the wheel, of strengthened metal rims in the development of the chariot, marked a turning point in the warfare of the Hyksos in Egypt, the Amorites in Mesopotamia, and the Hittites in Asia Minor. The development of new methods of warfare resulted in the emergence of the first fortifications; their development is documented in the first, oldest known (at the time of writing) settlements such as Harappa and Mohenjo Daro in the Indus Valley.

The first Hittite metallurgic centres in the southern coastal part of the Anatolian Plateau, from which metalwork spread to the European continent, were especially important. This process was linked to the development of more complex forms of social organisation, expressed in the dynamic growth of settlements, fortified settlements, and the emergence of the first city-states—which were locked in almost constant war with one another (Clark and Piggott 1965). The highest level of development of metal (copper, bronze, and iron) was, therefore, first reached in Southwest Asia. The application of metals, spread by Indo-European migrations, had massive significance for the development of tools and weapons, as well as the development of agriculture, new forms of warfare, and social organisation. The metal ages were especially important for Europe because the first historically documented mass migrations began in this era. The geographical core that played the role of the starting point of the largest migrations in the history of humanity thus far was the southern parts of the Eurasian steppe, especially the continuous lowland belt along the north coast of the Black Sea, the Caspian Sea, as well as Asia Minor, and the Iranian and Pakistani Plateaus.

These are areas in which different groups of inhabitants, which fall under the broad category of Indo-European peoples, lived. The oldest traces of Indo-European communities, primarily practicing pastoral economy, have been found east of the Dnjestar River (the core is located in contemporary Ukraine). There are numerous remains of horses, metal accoutrement for horses, and even traces of the ceremonial burial customs—tribal/familial leaders with their horses—as well as numerous skeletal remains of domesticated animals (mostly sheep, goats, cows, and pigs). A significant number of these people evidently died violent deaths and considerable number of ancient weapons have been found, which speaks to the way these warrior–herder groups lived. Additionally, the many remains of wheels and wagons (horse-drawn) that have been found speak to their great mobility. Archaeological research from the Dnjestar River all the way to the eastern edge of the European continent has confirmed the gradual expansion of these groups toward the west (Mallory 1989).

The causes that led to these great migratory movements are still a topic of scientific interest, and the causes that are most often mentioned are climatic, social, population, economic, and psychological. Although newer research especially emphasises climate disruptions, which could have negatively reflected on the lives of the people living in the area in question due to an increasingly dry era, other possible causes, particularly political ones, can still not be ruled out. Most scientists agree how the first groups of Indo-European people arrived in Europe in multiple waves in the second half of 3,000 BC. Many findings support this, such as the Assyrian texts found at Kârum Kaneš (Kültepe in modern

Turkey). Historical texts show how the era of 2300–1900 BC was characterised by expressed political instability, which followed continual changes in leadership in the large kingdoms in Egypt and Mesopotamia. This pushed the Indo-European tribes that lived on the edges of these great civilizations to migrate to the west (Clark and Piggott 1965).

These peoples primarily practiced pastoral economy, which was, along with the use of metal weapons, the main basis of social organisation. One of the key elements of their social organisation was their expressed warrior mentality based on the tribal system. Indo-European peoples were primarily familiar with extensive pastoralism and their land cultivation knowledge was less well-developed. As such, there was a constant need for large grazing areas, and conflict over living space was one of the main reasons for their frequent clashes. This meant that constant mobility, belligerence, the need to expand territory and control it, as well as social organisation based on a strict hierarchy and the prominent role of leadership, were the main characteristics of Indo-European peoples. Neolithic humans, with their pacifist mentality that was linked to their remarkable agricultural economy, sedentary lifestyle, and the role of women in their society (matriarchy[6]), could not effectively resist the Indo-European invasions. Thereby, nearly all traces of Neolithic society were obliterated in a very short period.

2.4 Environmental Priorities in the Aeneolithic—Mountains and Hillforts

New beacons of settlement in the metal ages developed in the Dinarides, which had had a mainly peripheral significance during the Neolithic. Numerous remains of material culture from this era bear evidence to this: first and foremost numerous fortifications (hillforts) that Indo-European settlers raised on the difficult-to-reach relief, which were ideal for defence and control of the surroundings (Rogić 1990). Additionally, the Dinarides boasted sites for iron mines and plentiful wood, needed to produce coal used in blacksmithing, as well as water resources that are the main metallurgic energy source.

The pastoral economy, developed over two millennia, had great consequences in terms of shaping the cultural landscape. In the metal ages, the Dinaric area was affected by degradation of vegetation, primarily forests, and the Pannonian Basin even more so. Analyses of animal remains in some archaeological localities in Istria from that era show how roughly half of the domestic animals consisted of sheep and goats, primarily the latter. A large number of items made from animal bone or related to animal husbandry (filters, roasters), as well as artistic depictions of the herding lifestyle, additionally confirm the domination of animal husbandry (Buršić-Matijašić 2008). Only later, during the Hallstatt period of the La Tène era (from 450 BC to Roman conquest), with the mixing of old Illyrian and Illyro-Celtic cultures, did farming increase due to the transition from migratory to sedentary land cultivation, primarily of grains (wheat, oats, barley, and millet), beans, peas, and grapes. Simultaneously, there was a transition from nomadic herding to transhumance, which was characterised by a seasonal vertical movement toward better pastures: toward the coast and Pannonian areas in the winter and into the mountains in the summer (Rogić 1956, 1990).

There is evidence which points to permanent settlement of some hillforts and their aspiration to overcome their function as temporary refuges to become real settlements. Findings also indicate that large hillforts might have had up to 1,000 inhabitants, the mid-sized 100–200, and the small around 50 (Buršić-Matijašić 2008). The Indo-European peoples who settled in this territory (Fig. 2.4) did not leave written evidence of their cultures, and, as such, clearly differentiating one from another remains a problem (e.g. Pre-Illyrians, Ur-Illyrians, Proto-Illyrians, Illyrians, Celts, and Illyro-Celts). The modern

[6]Evidence in support of the existence of Neolithic Matriarchy is also documented by numerous ceremonial figures depicting women.

Fig. 2.4 Illyrian and Illyro-Celtic tribes. *Source* Regan (2003)

understanding of these cultures come from sources in late Antiquity (ancient Greeks and Romans). On the basis of toponomastics, onomastics, and the accounts of chroniclers from Antiquity, it is thought today that Illyrians stem from groups that came from the Eurasian steppe and arrived in Southeast Europe at the end of the third millennium BC (Zaninović 1997).

Accounts from Antiquity describe the frequent warring among the established Illyrian, Pannonian, and Illyro-Celtic peoples, caused by the need for pastures, as animal husbandry was the dominant economic activity, upon which the existence of the tribe and its people depended. Certain accounts of Greco-Roman authors are of particular significance, such as Polybius, Appian, and Pliny the Elder, who lived and wrote during the transition from the old era to the new (Polybius 1999; Džino and Domić Kunić 2013), when the Romans organised the contemporary Croatian area into two provinces (*Dalmatia* and *Pannonia*). From their accounts, one can discern that the basic spatial unit was the village (*pagus*), which consisted of multiple smaller communities/hamlets (*vicus*) organised on the basis of blood relation. *Pagus* was understood, apart from its purely spatial nature, as a higher level of societal integration, and its centre was a permanent hillfort settlement (*castellum*).

The Romans had to put in significant effort to conquer some of these *castella*, so it can be concluded that they were well fortified and defended. The Romans also tended to destroy them after conquest, so that the conquered Illyrian tribes wouldn't re-fortify within their walls. Inhabitants of the *pagus* and their constituent *vici* constituted a community of a type of rural municipality (*conciliabulum*), which was led by a council of elders, and members of the council (*vicimagister*) were selected from some of the *vici*. The leader of the *pagus* itself was one of the distinguished leaders of the *vici* and bore the title *magister*. Rural municipalities were primarily integrated on the social level, and this integration also moved to the territorial level over time as a municipality grew to become proto-urban. Neighbouring rural municipalities negotiated communal use of pastures they shared, as this was often a driver of armed conflict (Suić 1976).

In the structure of the livestock pool, sheep and goats prevailed. Zooarchaeological analyses of animal remains in two locations in the area of Pula, where the Histri tribe lived, support this. They show how about half of the livestock pool consisted of sheep and goats (50.30%), while cattle made up a significantly smaller portion (29.60%), pigs still less (11.29%), and reindeer and animals for hunting made up the remainder (6.30%) (Matijašić 1998). The livestock pool distribution was similar among the Iapydes tribes in the region of Lika and Croatia's Dinaric mountainous areas[7] (Olujić 2007). Data on the number of hillforts, as relevant settlement indicators of individual parts of Croatian territory, were not completely reliable because only some of them were permanently settled. According to current understanding, in the area of the Istrian Peninsula, one hillfort had roughly 12 km^2 of the surrounding area. In contrast, hillforts on the island of Lošinj only had 2.5 km^2 of accompanying area. The total number of hillforts in a given area varied heavily. There were 436 established in Istria (the existence of 240 has been proven), 14 in the immediate area of Rijeka, 26 on the island Krk, 29 on the island Cres, 30 on the island Lošinj, 9 on the island Rab, around 400 in northern Dalmatia, 17 on the island Brač, 4 on the island Lastovo, 7 on the island Mljet, 57 in the Neretva Delta, and around 150 in Lika and Dinaric mountainous Croatia (Buršić-Matijašić 2008).

Conclusions can be made about the size of municipality territory based on the writings of the Roman (of Greek origin) historian Appian in the second century regarding the conflict between the Delmatae and Liburnian tribes over control of Promonom (the contemporary village of Tepljuh near Drniš, in the Dalmatian Hinterland), as well as writings regarding the cessation of disputes over municipality boundaries among neighbouring municipalities. It can be concluded that the radius of a rural municipality's gravitational area was roughly 10 km, when we also add in the average distance between their respective main centres of 10–15 km. This was a significant area during an era with weak transportation communication. Conversely, there are no traces of territorial municipalities on the islands, apart from on Krk, Cres, and Rab, because they were administered from the nearest settlement on the coast (Suić 1973).

It is easy to understand why the reconstruction of population density during the era of domination by Indo-European tribes in contemporary Croatian territory is a very difficult task. The common assessment is that roughly five inhabitants per km^2 were the maximum capacity of the territory, whereby it is important to note that settlement gradually increased toward floodplains in the Pannonian Basin and karst poljes, while mountainous areas—especially in karst regions—were only periodically used and were weakly settled at best (Rogić 1990).

The example of the Velebit coastal slope (Velebit Littoral) is very interesting because the boundaries between Iapydes tribal groups is preserved. These boundaries, as well as stone carvings, are from the Roman era. They clearly indicate the emergence of degradation of vegetation cover and problems with water supply. They also indicate problems with how the tribes themselves demarcated boundaries between smaller tribal communities (Rogić 1990, 2003), which was promoted by Roman leadership on the

[7] In Iapydes dwellings, numerous remains of ceramic objects used to process milk have been found.

basis of old pre-Roman demarcation practices. It is especially important to emphasise that, in this instance, one can see the traditional demarcation along the transversal of the Adriatic Sea–Lika Basin, which is predisposed to seasonal, vertical transitory movement, from pastures to the sea in the winter, and back to the rich pastures of Velebit and the Lika Basin in the summer (Rogić 1958).

References

Alexander J (1977) The "frontier" concept in prehistory: the end of the moving frontier. In: Megaw JVS (ed) Hunters, gatherers and first farmers beyond Europe—an archaeological survey. Leicester University Press, Leicester, pp 25–40

Baryshnikov GF, Markova AK (1992) Early man and mammals during the upper pleniglacial: main mammal assemblages between 24,000 and 12,000 yr. B. P. In: Frenzel B, Pécsi M, Velichko AA (eds) Atlas of paleoclimates and paleoenvironments of the northern hemisphere: late pleistocene-holocene. Geographical Research Institute, Hungarian Academy of Sciences, Gustav Fischer Verlag, Budapest-Stuttgart, pp 127–129

Bognar A, Faivre S (2006) Geomorphological traces of the younger pleistocene glaciation in the central part of the Velebit Mt. Hrvatski geografski glasnik 68 (2):19–30

Clark G, Piggott S (1965) Prehistoric societies. Hutchinson & co., London

Buršić-Matijašić K (2008) Gradinska naselja Istre – gradine Istre u vremenu i prostoru. Leykam International, Zagreb

Dugački V, Regan K (eds) (2018) Hrvatski povijesni atlas. Leksikografski zavod Miroslav Krleža, Zagreb

Džino D, Domić Kunić A (2013) Rimski ratovi u Iliriku – povijesni antinarativ. Školska knjiga, Zagreb

Glamuzina N, Fuerst-Bjeliš B (2015) Historijska geografija Hrvatske. Sveučilište u Splitu, Split

Jäger H (1969) Historische Geographie. Georg Westermann Verlag, Braunschweig

Leroi-Gourhan A (1983) Les chasseurs de la préhistoire. Éditions A.-M, Métailié, Paris

Mallory JP (1989) In search of the Indoeuropeans—language, archaeology and myth. Thames & Hudson, London

Matijašić R (1998) Gospodarstvo antičke Istre – arheološki ostaci kao izvor za poznavanje društveno-gospodarskih odnosa u Istri u antici (I. st. pr. Kr. – III. st. posl. Kr.). Zavičajna naklada „Žakan Juri", Pula

Novak G (1944) Prošlost Dalmacije, knjiga prva: od najstarijih vremena do Kandijskoga rata. Hrvatski izdavalački bibliografski zavod, Zagreb

Olujić B (2007) Povijest Japoda. Srednja Europa, Zagreb

Orme B (1977) The advantages of agriculture. In: Megaw JVS (ed) Hunters, gatherers and first farmers beyond Europe—an archaeological survey. Leicester University Press, Leicester, pp 41–49

Polybius (1999) Istoriai—the histories. Harvard University Press, Cambridge

Regan K (ed) (2003) Hrvatski povijesni atlas. Leksikografski zavod Miroslav Krleža, Zagreb

Rogić V (1956) Razlike pejzaža velebitskih padina. Geografski glasnik 18:15–32

Rogić V (1958) Velebitska primorska padina – prilog poznavanju evolucije krškog pejzaža (nastavak). Geografski glasnik 20:53–110

Rogić V (1990) Regionalna geografija Jugoslavije, knjiga 1 – prirodna osnova i historijska geografija. Školska knjiga, Zagreb

Rogić V (2003) Sveti Juraj – mikrogeopolitički retrospekt i prospekt. In: Babić S (ed) Spomenica Svetog Jurja. Udruga za zaštitu i promicanje razvitka Svetog Jurja, Rijeka, pp 9–27

Suić M (1973) Odnosi grad-selo u antici na istočnoj jadranskoj obali. Jugoslovenski istorijski časopis 3–4:13–40

Suić M (1976) Antički grad na istočnom Jadranu. Sveučilišna naklada Liber, Zagreb

Šegota T (1982) Razina mora i vertikalno gibanje dna Jadranskog mora od ris-virmskog interglacijala do danas. Geološki vjesnik 35:93–109

Šegota T, Filipčić A (1996) Klimatologija za geografe. Školska knjiga, Zagreb

Zaninović M (1997) Iliri u antici na hrvatskom tlu. In: Supičić I (ed) Hrvatska i Europa – kultura, znanost i umjetnost: svezak I. Srednji vijek (VII – XI. stoljeće) rano doba hrvatske kulture. AGM, Zagreb, pp 55–71

3 Antiquity—The First Model of Territorial Organisation

Abstract

Classical times represent a major turning point in the historical–geographical development of the territory that would become Croatia. Thanks to two great civilizations that existed in close proximity—the ancient Greeks and Romans—contemporary Croatian territory became a highly developed cultural landscape. Many traces of these giants of Antiquity remain to present day, mostly on the islands and coastline of Adriatic Croatia. Ancient Greek civilisation is known to have been primarily a maritime culture, and Greek influence spread north into the Adriatic along with their ships. However, it did not leave deep marks on the cultural landscape. The core and extent of the influence of the Greek state organisation in contemporary Croatia were located on the islands of central and southern Dalmatia, which made up the basis of the cultural landscape in terms of new settlements—*poleis* (e.g. *Issa* and *Pharos*). The Greeks also introduced advanced Mediterranean agriculture, namely the cultivation of grapes. In contrast, the significance of Roman influence lies in the fact that it marked the first time in history that the entirety of contemporary Croatian territory would be incorporated into a unified and firmly organised political whole. The organisation of newly conquered areas into the administrative wholes of *provinciae*, with a clear system of subordination, marked the transition from a fragmented network of tribes to unity under a single state organisation. Contemporary Croatian territory was organised in two *provinciae*—*Pannonia* and *Dalmatia*—with an advanced urban and transportation network (rivers and roads).

Keywords

Croatia · Antiquity · Classical era · Ancient Greece · Ancient Rome · Poleis · Provinciae · Pannonia · Dalmatia · Mediterranean agriculture · Ager

Classical times represent a major turning point in the historical–geographical development of the territory that would become Croatia. Thanks to two great civilizations that existed in close proximity—the ancient Greeks and Romans—contemporary Croatian territory enjoyed a new, highly developed cultural landscape. Many traces of these giants of Antiquity remain to present day, dispelling some of the darkness that shrouds events of the distant past. Of particular interest are the numerous writings which these two great civilisations left behind, which mark the entrance of the eastern coast of the Adriatic into the historical era.

B. Fuerst-Bjeliš and N. Glamuzina, *The Historical Geography of Croatia*,
Historical Geography and Geosciences, https://doi.org/10.1007/978-3-030-68433-4_3

Stari Grad Plain, Hvar island. UNESCO World Heritage site
Stari Grad Plain represents a system of land use, established by the ancient Greeks in the fourth century BC. The Greek cadastral system has been fully respected during the continuous agricultural use of the plain over the last 2,400 years. The structure of the stone walls are the authentic Greek foundations (Photograph: Ivo Biočina. Reproduced with permission of the Croatian National Tourist Board)

3.1 The Spatial Impact of Ancient Greek Colonisation

The first great socio-political organisation in the Mediterranean arose in the Aegean Basin, not too far away from the land we call Croatia today. At its greatest level of influence, it controlled parts of Macedonia[1] and Albania, and also had direct land borders with Illyrian tribes. Although there was a healthy level of trade, the main ancient Greek cultural influence came to the area by sea, from the Aegean Basin. Namely, ancient Greek civilisation is known to have been primarily a maritime culture, and their influence spread north into the Adriatic along with their ships. This has been confirmed by Greek weaponry during 500–700 BC, found on the island Brač and in Bosnia, which likely came to the area by way of maritime trade (Rogić 1990; Novak 1944).

Additionally, remains of Corinthian ceramics found on the southern Adriatic islands Korčula and Vis, as well as near the littoral cities of Solin and Zadar, provide evidence of trade between Corinthian merchants and Illyrian tribes (Zaninović 1997). The causes of Greek expansion during Antiquity are found primarily in the geographical characteristics of Greece itself, which is largely a karst landscape with weakly developed hydrographic networks and a limited quantity of fertile soil. Ancient Greeks became (primarily sea-going) merchants due to the fact that the potential of agriculture for producing food was less significant. Furthermore, trade was not sufficient for subsistence during periods of population expansion. The solution in the end was that people left Greece and founded colonies in other parts of the Mediterranean, from the Black Sea coast to the Apennine Peninsula. The northern orientation of ancient Greek colonial expansion, toward the Ionian Sea and the southern tip of the Apennine Peninsula, was historically and geographically important for the development of contemporary Croatian territory.

In the Ionian Basin, colonists from various Greek *poleis* settled on Sicily and the southern part of the Apennine Peninsula and there they formed the first colonies with a significant level of cultural development. The centre of the colonies was *Syrákousai*, contemporary Syracuse on Sicily. Further waves of colonists began to enter the Adriatic Sea and established cities on its shores. They even made it to the contemporary Umbria Region in Italy, and also traded at the port cities of Spina and Adria in the farthest northern reaches of the Adriatic. Moreover, it was in this very area that mainland trade routes of the day happened to come into contact with their Mediterranean counterparts[2] (Rogić 1969; Kuntić-Makvić 1997).

The first significant marks on the Adriatic coast were left by Dionysius the Elder (Dionysius I of Syracuse), who ruled Sicily and the southern part of the Apennine Peninsula in the second half of 400 BC. He made Syracuse into a great naval power in the central Mediterranean by strengthening its fleet with hundreds of new ships. He controlled maritime traffic from Carthage and Sardinia in the west, to the Ionian Coast of Greece and the Strait of Otranto in the east, and all the way to North Africa (Matijašić 2009). Dionysius the Elder also certainly reduced the prior influence of Liburnia, one of the most developed Illyrian tribal communities—to which the term "Liburnian Thalassocracy" is incorrectly attributed (Rogić 1990).

Thanks to superior sea power, he extended the influence of the Syracusan Thalassocracy into the Adriatic part of the Mediterranean which also resulted in the founding of new *poleis*. The first among them was *Issa*, founded in 397 BC on the island Vis, where the town Vis stands today. Thereafter, other settlements on the nearby islands Hvar and Korčula were founded: *Pháros* and *Dimos* or *Dimale* (on Hvar), and *Kórkyra'e Mélaina* and Lumbarda (on Korčula). These new settlements and the spread of advanced

[1]The territory of the contemporary Republic of North Macedonia.

[2]Regarding the significance of *Adria* as a centre of international trade, the Adriatic Sea itself derives its name from this city, as the ancient Greeks referred to the northern part of the Adriatic Sea as the "Adriatic Bay" (*Adrías kólpos*).

Mediterranean agriculture (especially cultivation of grapes, olives, figs, and grains) created a new cultural landscape in the islands of the southern Adriatic.

The topographical position of most ancient Greek colonies was a result of the most basic need: security. Due to this concern, settlements were raised on peninsulas or islands located near the mainland. Close proximity to a natural harbour was also important, one that could accommodate ships and let the settlement take part in trade along the Adriatic coast. *Poleis* were built in areas with significant concentrations of fertile soil, so the colony could satisfy its basic need for food (Matijašić 2009).

The geographical core of the state organisation of the ancient Greeks in contemporary Croatian territory was located in the islands of central and southern Dalmatia. The reason for this is found not only in geographical factors, i.e. the close proximity of Magna Graecia in the "heel" of the Apennine "boot", but also in the fact that their superior sea power enabled easier defense and control of maritime lanes. The significance of the island Vis, from which sea travel toward the northern part of the Adriatic can be controlled, confirms this. It is in this very area that the influence of the Greek community on the original Illyrian inhabitants is most expressed, and led to the two peoples mixing. Names on graves of citizens of *Issa* as well as names on the Lumbarda Psephisma—an inscription carved in stone which speaks about the establishment of settlements on Korčula—bear witness to this (Kuntić-Makvić 1997).

Later, on the eastern Adriatic coast, several trade emporiums were founded: *Tragourion* (Trogir), *Epetion* (Stobreč near Split), and ancient Greek *Salona* near a pre-existing Illyrian centre (near Solin in the vicinity of Split). From this position in central Dalmatia, they could (and did) penetrate into the hinterland through the mountain pass at Klis (inland of Split) to strengthen their trade activities. In the Neretva Delta, near Illyrian *Narona*, they would build their own trading settlement. As opposed to the island chain, which had the role of the centre under Greek rule, the cities on the central Dalmatian coast were located in areas which were still controlled by Illyrian tribes. Therefore, they had the role of trading posts and were of secondary significance. This fact is additionally confirmed by the location of Greek trading settlements on communication routes with the continental interior, such as Kaštela Bay near Split in central Dalmatia and the Neretva River Delta in southern Dalmatia (Rogić 1990; Matijašić 2009; Džino and Domić Kunić 2013).

Because the ancient Greeks came to the Adriatic coast primarily as traders, rather than conquerors, and were in search of good agricultural land, the founding of the settlements themselves indicates the significance of the central Dalmatian coast in Illyrian society. The very reasons for this, and the very limited settlement of peripheral coastal parts of Croatian territory today, are the reasons why ancient Greek civilisation did not leave deeper marks on the cultural landscape. Furthermore, many of the marks that they did leave have been erased over the millennia. Some significant traces of ancient Greek land parcelling (cadastre) in the Stari Grad plain on Hvar (UNESCO World Heritage site) and the area of Vis (*Issa*) have been preserved. Cities took up little area themselves, and archaeological research has shown that *Issa* was spatially the largest (8.5 ha), and had the largest *ager* at its disposal (at least 450 ha). Conversely, coastal settlements were more compact and took up much less space, as the example of *Tragourion*, which took up only three ha, shows well (Rogić 1990; Matijašić 2009).

3.2 Ancient Roman Civilization as a Turning Point in the Formation of the Cultural Landscape

Due to significant differences in the level of development between the indigenous Illyrian social organisations and the Greek settlers, Illyrian tribes were not able to completely implement all the innovations arriving in their land at the time. This would only happen with the coming of the Romans, whose administration had much greater significance and left a greater mark in the cultural landscape, one that is still visible today.

The significance of Roman antiquity is twofold. First, it is important because this marked the first time in history that the entirety of contemporary Croatian territory was incorporated into a unified and firmly organised political whole. As opposed to the ancient Greeks, who settled the eastern shores of the Adriatic primarily for mercantile, maritime, and agricultural reasons, the Romans came as an exceptionally organised and powerful state organisation, looking to exploit the region's economic resources and to settle the area with Roman citizens. One of the other important reasons why the ancient Romans came to the area was the clear goal of subjugating the native Illyrian tribes and Illyrian states, which were important political entities in the Adriatic Basin. Thus, the eastern shore of the Adriatic and its entire hinterland were swallowed by the rising power of the Roman Republic. In order to realise its goals, the Romans emphasised well-arranged spatial organisation in the territories that they conquered. The basic elements of the aforementioned spatial organisation were (Rogić 1990):

- organisation of urban networks
- construction of a transit system
- administrative–territorial organisation of the area.

These three elements are causally linked, so it is very difficult to analyse them individually and ignore their interrelation, stemming from the beginning of Roman antiquity. Their respective development, however, was not totally synchronised. The ancient Roman state developed the elements that were most needed for its purposes at a given time (Rogić 1990).

In the conquering phase, the realisation of strategic and political goals took priority. Accordingly, the construction of a transit system to enable Rome's armies to move quickly was at the top of the list. For this reason, infrastructure works were begun—general road-building and repair of river canals and channels, essential for wagon transport and advanced forms of transport, i.e. primarily military logistics, wood, stone, and agricultural products. Control of existing eastern Adriatic harbours and the creation of new ones were essential to controlling the Adriatic and eliminating Illyrian pirate strongholds. Namely, piracy increased greatly following the founding of the first Greek colonies on the islands Vis and Hvar, which were bearers of rapid maritime trade development (Džino and Domić Kunić 2013).

Following strategic goals, which never completely lost their significance (especially during late Antiquity when the region came under constant attack by "barbarian" peoples), economic goals were pursued. The Romans enacted carefully considered economic policies on areas that they conquered. The lion's share of emphasis was placed on agrarian policy, as agriculture was of paramount importance. The first step was the confiscation of the most valuable agricultural land, by which the native population was forced further into the mountainous interior.[3] Closely linked to the first step was the second: the policy of colonising Roman citizens from the Apennine Peninsula, who were meant to be the backbone of the new government, as well as the instrument that would bring more advanced forms of economics to this new part of their domain. The organisation of newly conquered areas into the administrative wholes of *provinciae*, with a clear

[3]Historical sources of Antiquity, such as Strabo's *Geography*, which was written during the very period that the territory we call Croatia was being conquered, and Ptolemy's *Geography* from the second century AD, wrote about the implementation of such policies on the eastern Adriatic coast.

system of subordination, symbolically and truly marked the transition from the fragmented network of tribes to unity under a single state organisation.

3.2.1 The Urban Network as the Main Bearer of Spatial Organisation

The urban network was the first important element which the ancient Roman state organisation built into the landscape and spatial–functional organisation of a newly conquered territory (Rogić 1990). Although urban settlements, mostly of the proto-urban type, had already sprung up in contemporary Croatia, within the framework of Illyrian and ancient Greek civilisations, under the ancient Roman state organisation, they would achieve new heights of development. Namely, Roman civilization was primarily urban and cities were crucial to spatial organisation. They had many simultaneous roles, from administrative and economic to military, religious, and cultural. Cities were meant for the role of politically and economically integrating a given area. The amount of effort the ancient Romans put into the development of urban networks and in each individual city is understandable, given their role. The Romans were practical and in areas that already had developed networks of cities (such as the eastern Adriatic coast), they simply improved them and built new cities in areas that lacked them (e.g. the Pannonian Basin).

The Roman urban network was organised hierarchically, as all urban settlements can not have the same importance in a given space, and in the administrative and political division. Settlements of the urban type could thus have an administrative status of a higher (*coloniae*) or lower rank (*municipium*). Cities with the status of a colony had a central role in the space. They were population spreading points for settlers from the Apennine Peninsula and other parts of the Republic, as well as economic and cultural incubators that spread innovation to their surroundings. The domination of coastal cities shows the Adriatic orientation of the area during this era. The urban centres in the Pannonian Basin, however, were very highly developed and were economically and strategically more important than coastal cities. Cities in the Pannonian Basin rose to dominance during Diocletian's reforms (at the end of the third century) when new political and economic circumstances necessitated changes in the relationships within urban networks (Rogić 1990; Suić 1997). The causes can be found in the economic structure of ancient Roman civilization itself, which put primary emphasis on crop growing.

3.2.1.1 Town and *ager*—The Functional Whole

Due to the development of crop cultivation in the surroundings of cities, ancient Romans systematically planned out cadastral parcellation (Rogić 1990; Suić 1976) of the most valuable agrarian land (*ager*). Namely, the Roman state claimed the right to unilaterally decide how to use the territory that it conquered, especially those conquered by force, as was the case in contemporary Croatia. Because of this, shortly after the installation of Roman leadership, all agricultural land was confiscated. There were different categories of *ager*, which changed depending on the era of development of the Roman state. During the transition from republic to empire, when contemporary Croatian territory came directly under Roman control, agricultural land was divided into three categories as follows.

Ager publicus was the term for land owned by the state, and encompassed the most fertile and valuable parcels of land, which were measured in detail, bounded, and visibly marked. The Roman state usually handed their administration over to local municipal communities in exchange for paying a tax (or in exchange for a portion of the yield in some cases). Local authorities further divided (via centuriation) *ager publicus* into smaller parcels which were leased to members of the Roman aristocracy and richer citizens. Those who leased these lands had to maximise their output and earnings, because they had to regularly pay a lease fee. Unsurprisingly, only the most valuable crops were sown on these parcels of land, such as grapevines, olives, figs, and various grains (primarily wheat, and also barley, rice, and oats).

Ager privatus was the term for land owned privately by Roman citizens. The Roman state gave this land to local municipal communities and was not involved with it after that. *Ager privatus* was typically under the administration of cities that functioned as local centres. It encompassed lands of lesser quality, which the Roman state showed no interest in exploiting and thus made them available to the native community without changing the existing ownership relations. Therefore, parcels of land might be communally owned by multiple rural settlements or owned by an individual who was able to make them available for use without restriction. In contrast to state-owned lands, *ager privatus* was not measured and bounded in detail, and its boundaries were typically natural elements (streams, canals, hills, forest vegetation, field paths, etc.).

Ager compascuus encompassed areas intended for animal husbandry, forests, and non-arable land. These lands, which were sometimes referred to as *compascua et silvae* (pastures and forests), maintained existing property status, and local communities were responsible for them, regardless of whether they were made up of native population or Roman citizens. *Ager compascuus* had the status of public good and was owned by municipal communities in perpetuity, and all of a given community's inhabitants were equally able to exploit it, for needs both public and private.

Measured and parcelled *agri* were divided into smaller, square-shaped parcels called *centuriae* (710 m long on each side). The policy of shaping *ager* in contemporary Croatia was completely undertaken in areas with the most fertile and arable land, whereby said land mostly went to members of the land-owning aristocracy from the Apennine Peninsula. These aristocrats organised extensive agricultural estates (*latifundium*) in excess of 500 juger (1 *jugerum* = 0,26 ha) located in the Pannonian Basin and the most agriculturally valuable parts of the eastern Adriatic coast (Istria, Ravni Kotari, and Neretva Delta). The majority of those working these lands were settlers from the Apennine Peninsula.

In areas of large cities, *agri* grew quite large, such that one *ager* of the city of Pula (1,100 *centuriae*) covered around 50,000 ha, and one ager of the city of Poreč (700 *centuriae*) around 35,000 ha. The average size of land possessed in areas of *ager* of the aforementioned Istrian cities ranged from, according to current reckoning, 359 to 633 juger, and were in the category of mid to large-sized *latifundia*. The average agrarian density was very rational: an average of 105 inhabitants on a holding of 600 juger, or 0.67 inhabitants/ha (Matijašić 1988; Suić 1976).

The Roman colonial population consisted of settlers from the Apennine Peninsula and assimilated, romanised indigenous people. Settlers from the Apennine Peninsula belonged to the group of small landowners—peasants who couldn't compete with the cheaper agricultural products from *latifundia* that were worked by slave labour. Colonists were very familiar with advanced agricultural techniques, such as crop rotation, fertiliser, advanced tools and techniques for cultivating land and individual crops, as well as the use of fallow land; and they played an important role in the spread of modern agriculture to the eastern Adriatic coast and the Pannonian Basin.

The role of indigenous farmers was completely marginalised. They continued to work the same land that they always had. The only change was that they had to pay dues to its new owners: Roman aristocrats. Although *ager publicus* had great commercial significance, *ager privatus* played a larger role from a societal viewpoint. Namely, the institution of private ownership, which had secondary significance to Illyrian and Pannonian peoples, initiated the process of societal division by classes and the end of traditional societal relations. Moreover, *ager compascuus* had the greatest significance to autochthonic inhabitants because they traditionally practiced animal husbandry as a primary economic activity. Additionally, coastal and Dinaric areas of Croatia, as well as individual areas in the Pannonian Basin (apart from the eastern parts and lowlands along large rivers), offered excellent conditions for animal husbandry. Roman authorities divided pastures, which had to that point been used jointly by neighbouring communities, and placed them under the direct control of cities to which the respective *ager* belonged.

This further damaged the societal structure of autochthonous peoples, who did not understand the strictly determined borders between land parcels. In place of neighbouring communities, indigenous folk reoriented toward cities, which became new social organisation centres. Additionally, the main urban centres, which administered the *agri*, prevented further development of indigenous settlements and their urbanisation. The example of Istria, where the land was under the control of *Tergestum* (Trieste), *Pola*[4] (Pula), and *Parentium* (Poreč), shows this clearly. The development of all other centres, apart from *Caprisa* (Koper), in the area of these three cities' *agri* was prevented. In Ravni Kotari the situation was the same, and the area was controlled from *Iader* (Zadar). The story is the same in central Dalmatia and the Neretva Delta, which were controlled from *Salona* (Solin) and *Narona* (Vid in the Neretva Delta), respectively. Conversely, the development of autochthonous settlements and their transformation into smaller, proto-urban centres was most dynamic in regions with small areas of *ager publicus*, such as Bukovica in the hinterland of Zadar, the coastal area below Velebit Mountain, Kvarner islands, and the Istrian coast up to the Raša River (Suić 1976).

Archaeological research has shown that various parts of given *agri* were not equally exploited. The best example of this are the *agri* of *Pola* and *Parentium*. The highest levels of exploitation, as well as settlement, were concentrated in the western parts of the *agri* in areas close to urban settlements. Settlement density dropped dramatically to the north and east—toward the interior of the peninsula—i.e. areas where indigenous herders dominated. Such division was also linked to geographical characteristics of the terrain, as well as with the main roads that connected the main urban settlements along the western coast of Istria (Matijašić 1988).

A new element in the landscape that represented a new type of rural settlement emerged in the Roman era. It was called *villa rustica*. They were raised on state lands in the vicinity of cities by rich citizens or members of the agricultural aristocracy. *Villae rusticae* were complex wholes, oriented toward the cultivation of individual crops, processing agricultural products, producing and storing food products (oil, wine, flour, etc.), and agricultural tools and equipment production. Some of them also had residential characteristics, along with their productive characteristics. This is shown in the villa in Verige Bay in the Brijuni island chain off the western coast of Istria (Suić 1976). A great number of such objects in Istria indicates peaceful development, economic prosperity, and also that the indigenous Histri tribe had been broken (during the second century BC) and forced deep into the interior of the peninsula, to the mountainous areas of Ćićarija and Učka.

A very high level of agricultural production was achieved in the Pannonian Basin in Antiquity, primarily of grains that were exported via northern Adriatic harbours. Due to this, cities in the basin enjoyed exceptionally dynamic development. Both the fertile soil and the rich forests were bases of the strong development of these cities, and encouraged Roman citizens from the Apennine Peninsula to resettle in the area. An additional stimulant for their growth was the considerable Roman military presence, charged with guarding the borders of Rome's domain, who needed a constant supply of various products. The Roman interest in exploiting the Pannonian area was considerable, and can be seen in the great efforts that Romans made in clearing forests, draining swamps, and shaping river channels. They had already significantly changed the natural landscape of Pannonia by the second century AD.

The shaping of the new cultural landscape was most expressed near large urban centres, which is still visible in the landscape to the time of writing. Autochthonous Illyrian herders were

[4]The full official name was *Colonia Pietas Iulia Pola*.

forced from the most valuable agricultural lands into the less-valuable mountainous belt. This additionally increased landscape degradation in the coastal belt.

3.2.2 The Transportation System—A Crucial Element of Spatial Organisation and Transformation

The transport system was the second important element of territorial organisation and the cultural landscape. Roads linked the main urban centres to each other. The Romans demonstrated their superior power, in relation to Illyrian–Celtic and ancient Greek civilizations, in the area of road building itself. Road building and maintenance was a top priority of the Roman state. Although they were primarily built for administrative and military purposes, they were also used extensively by merchants and traders. There were rest stops along roads (*mansiones*) and places to change horses (*mutationes*). Roman army units ensured safety and security along the roads. The network of roads in contemporary Croatian territory linked it to the rest of the empire, primarily with the core in the Apennine Peninsula. Roman roads typically used existing prehistoric trails and paths along the most desirable natural routes. In Croatian territory, we can distinguish longitudinal from transverse roads (Fig. 3.1).

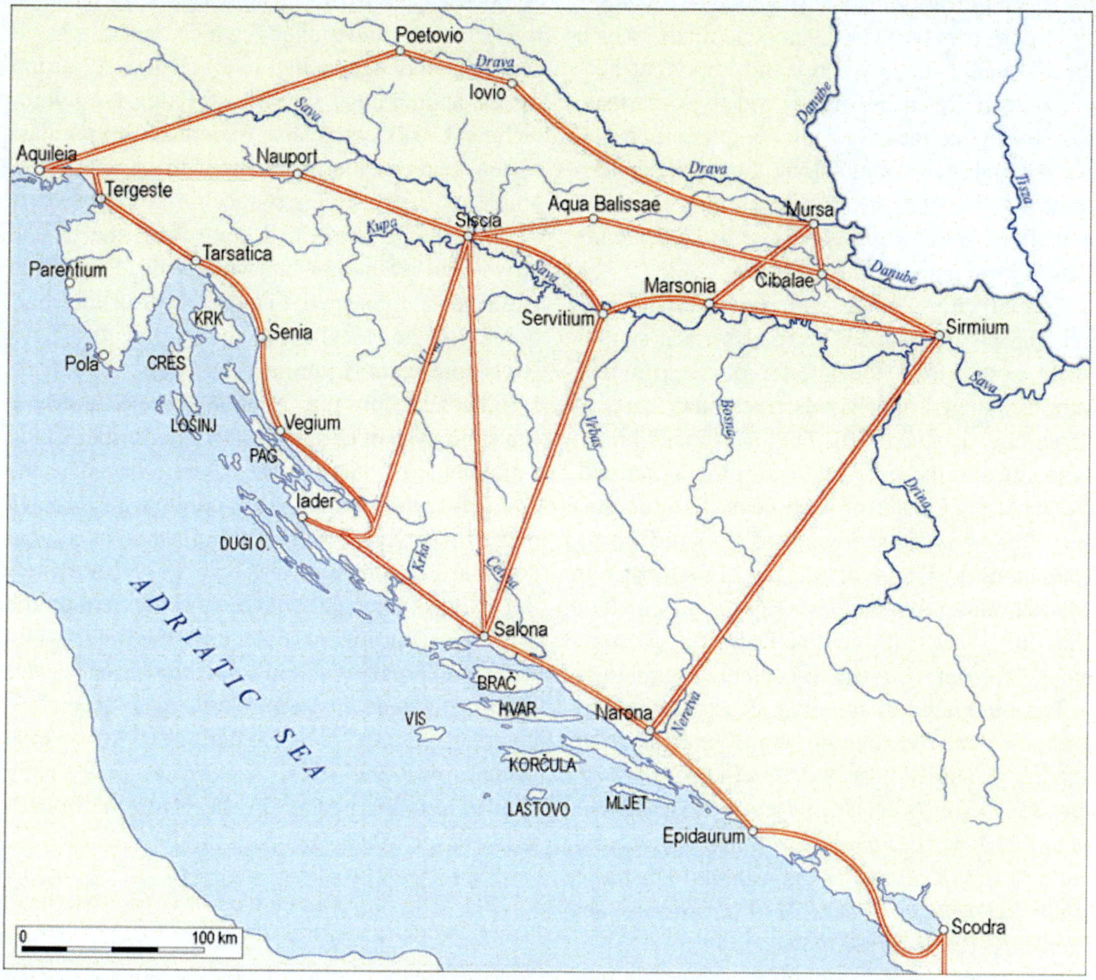

Fig. 3.1 Main Roman road transport and urban networks

Of the aforementioned longitudinal roads, the one that led from the far eastern parts of the empire toward the Apennine Peninsula played a key role. It linked *Sirmium* (Srijemska Mitrovica in Serbia), *Marsonia* (Slavonski Brod on the Sava River)—with a detour to *Cibalae* (Vinkovci) and *Mursa* (Osijek on the Drava River)—with *Siscia* (Sisak on the Sava River), *Nauport* (Novo Mesto in Slovenia), and *Aquileia* on the far northern coast of the Adriatic. A second longitudinal roadway also linked other large urban centres, extending along the northern Drava River corridor through the centres *Poetovio* (Ptuj in Slovenia), *Iovia* (Ludbreg), *Mursa*, and *Cibalae*. These two main roads were linked to one another by a series of perpendicular roads, of which the most important was the diagonal road *Siscia–Aqua Balissae* (Daruvar)–*Cibalae*.

A dense network of smaller roads which linked smaller urban centres and important economic areas (quarries, mines, and river marinas) was also maintained. The most important coastal longitudinal road wound along the Adriatic coast from *Aquileia* toward *Iader*, *Salona*, *Narona*, onward to *Scodra* (Shkodër) and *Dyrrachium* (Durrës) in contemporary Albania.

Coastal cities, along with harbour, military, and administrative roles, were also trade centres where goods from the interior of the province were exchanged with goods from other parts of the empire (Rogić 1990). The main roads linked these cities to the main centres in the hinterland. The most significant of such centres in the area were *Tarsatica* (Trsat—a part of modern Rijeka), *Senia* (Senj), *Vegia* or *Vegium* (Karlobag) in Kvarner Bay, *Aenona* (Nin), *Iader*, *Scardona* (Skradin), *Salona* in central Dalmatia, *Narona*, and *Epidaurum* (Cavtat) in southern Dalmatia.

The hinterland of the main coastal cities was marked by the dynamic growth of smaller urban centres, primarily near *Scardona* (*Arausa*, *Burnum*, *Loranum*, *Ninia*, *Promona*, and *Synodium*), *Iader* (*Asseria*, *Blandona*, *Corinium*, *Clambetis*, *Hadra*, and *Nedinum*), and *Salona* (*Aequum*, *Bilubium*, *Nouae*, *Rider*, *Setovia*, *Siculi*, and *Tilurium*) (Rogić 1990; Novak 1944).

In contrast, the most important transverse road linking the Pannonian Basin and the eastern coast of the Adriatic was that which lead to *Salona* from *Siscia* and *Servitium* (Bosanska Gradiška) on the Sava River. The roadway toward *Narona* from *Sirmium* that traversed the valleys of the Drina, Bosna, and Neretva rivers, through the Dinarides, was of secondary importance. The third transverse road was *Siscia–Iader* (and on to *Scardona*), following the Una River to the middle of the Lika Region, where the road branched toward harbours in Kvarner Bay (*Senia* and *Tarsatica*).

Sea and river traffic had an equal (or perhaps greater) level of importance than road traffic. It was the primary method of transporting construction materials and agricultural products (especially grain, oil, and wine), and also served to move the might of the legions. For this reason, comprehensive hydro-technical projects on large rivers were undertaken, which encompassed cleaning and deepening river channels, raising dykes, and creating *kopitnica*.[5] Extensive hydro-technical works were also undertaken to regulate certain parts of the channels of the Kupa, Sava, Vuka and Bosut rivers (Gračanin 2011).

Three large rivers had key roles—the Drava, Sava, and Danube—by which the Pannonian Basin was connected to northern Adriatic harbours (Rogić 1990). The course of the Sava River to *Emona* (Ljubljana) had key geostrategic significance for the Pannonia–north Adriatic transport system (Rogić 1965). The former Celtic settlement of *Segestica*, which became the Roman city of *Siscia*, at the conflux of the Sava and Kupa rivers, had key significance as a large colonial Pannonian city (Rogić 1977). *Andautonia* (at the site of Šćitarjevo—a settlement on the Sava River located near Zagreb), the predecessor to contemporary Croatia's leading urban centre Zagreb, had an important role as a secondary centre. The city was located next to a stone bridge over the river, and there were many smaller satellite centres in the nearby surroundings (Rogić 1990; Gračanin 2012).

[5]Dykes that were raised along river banks, intended for horses towing boats upstream.

3.2.3 Administrative Organisation as an Element of Spatial Transformation

The administrative–territorial organisation of the area changed throughout centuries of rule under the Roman Empire. For much of this time period, however, the administration was divided into four Pannonian and two Adriatic provinces, which made up the greater province *Illyricum*. It is important to emphasise that, within *Illyricum*, Pannonian provinces had greater significance due to their economic strength, and the strategic/political importance that came with their proximity to the hard border (*limes*) of the Roman Empire, i.e. the Danube River.

The significance of the Pannonian area increased steadily, which is demonstrated by its continual division. The province was first formally organised in the first century, then split into two provinces in the second century, and each of those was split again in the third century, making 4 provinces (Fig. 3.2). Such administrative divisions over time are indicators of an increasing need for stronger oversight of the defense of the Pannonian area, as well as a more efficient administrative function under conditions of increasing settlement density and economic development. During the Roman era, the weight of settlement shifted from the mountainous areas to the lowlands, most likely alongside population advances; we can thereby assume a maximum density of 10 inhabitants/km^2 (Rogić 1990).

Within the Pannonian Basin, the Sava traffic corridor was crucially important. It connected the northern part of the so-called 10th region populated by ancient Illyrian peoples of the northern Adriatic—Veneti and Histri—to the Danube. Roads were built in the corridor and the Sava River channel was worked over for river traffic. Three large urban centres were established in the area: *Siscia*, *Marsonia*, and *Cibalae*. The Drava Corridor, upon which the centres of *Mursa*, and (on the right bank of the Danube) *Ad Militares* (Batina), *Ad Novas* (Zmajevac), *Teutoburgium* (Dalj), and *Cornacum* (Sotin) were founded, was of secondary importance.

In the second century, *Pannonia* was split into two smaller provinces, which corresponded to large spatial wholes: *Pannonia Superior* and *Pannonia Inferior*. The former encompassed the peri-Pannonian (western) part of contemporary Croatian territory, and the latter the proper Pannonian region in the east (eastern Slavonia, Baranja, and Srijem). At the end of the third century, these two provinces were further divided, resulting in four administrative units. The first of these new units was *Pannonia Savia*, which encompassed the entire contemporary Croatian peri-Pannonian area to the Drava in the north. The administrative and primary centre of the peri-Pannonia region was *Siscia*. Complex hydro-technical regulation was necessary to build the city, located in the flood-prone and swampy conflux of the Kupa and Sava rivers, indicating the strategic importance of its location. *Siscia* took the role of primary transit node in the Sava Corridor, and simultaneously maintained a link with the Dinarides that were rich in mines, as well as with the Sava–Danube node *Singidunum* (Beograd) (Rogić 1990).

The far northwestern part of Croatia (Međimurje and Podravina regions) was annexed into a province called *Pannonia Prima*, in which *Savaria* was the main centre (Szombathely, Hungary). *Pannonia Secunda* encompassed the proper area of Pannonia, in which *Sirmium* was the main centre. This urban centre, due to its location in the Sava Corridor and proximity to the Danube, experienced strong development. The economic importance of the strong agricultural base as well as the transverse roadway links with Dalmatian harbours on the Adriatic (*Salona*) are other factors that played into the strong development of the area. It is very important to note that it was in the era of late Antiquity, during Diocletian's reforms, when *Sirmium* became the administrative centre of the *praefectura praetorio per Illyricum* (praetorian prefecture of Illyricum), which united the Pannonian, Adriatic, Dinaric, and parts of the Ionian–Aegean areas (Rogić 1977).

The secondary centres of the province were *Cibalae*, the main transit node in the province

Fig. 3.2 Administrative organisation of the Roman Empire at the turn of the third century to the fourth century (*Source* Dugački and Regan 2018)

that linked the Sava and Drava corridors with the Danube, and *Mursa* which had the greatest strategic significance due to its proximity to the conflux of the Drava and the Danube. The fourth province was *Pannonia Valeria*, which included the contemporary Baranja Region in the far northeastern part of Croatia. The main centre of this province was *Sopianae* (Pécs, Hungary).

In contrast, the coastal and mountainous parts of contemporary Croatia were organised into two administrative wholes. During the transition from the old era to the new, the majority of Istria was annexed directly to the 10th region of the Italian part of the Empire. The main administrative centres in this part of the Roman Empire were *Aquileia*, *Tergestum*, *Pola*, and *Parentium*. They spread Romanisation, culture, and involved the autochthonous Histri in Roman political and socio-economic organisations. Based on the analyses of epigraphic monuments, we can conclude that the Roman ethnic element dominated in areas of *agri*, and was almost not present outside of them due to the dominance of the native Histri. In the area on the edges of *agri* in the interior of Istria, however, we find the most expressed romanised Histri communities.

Istria was already at a higher level of economic development in relation to other parts of the eastern Adriatic coast. The weight of economic activity and population in Istria was in its

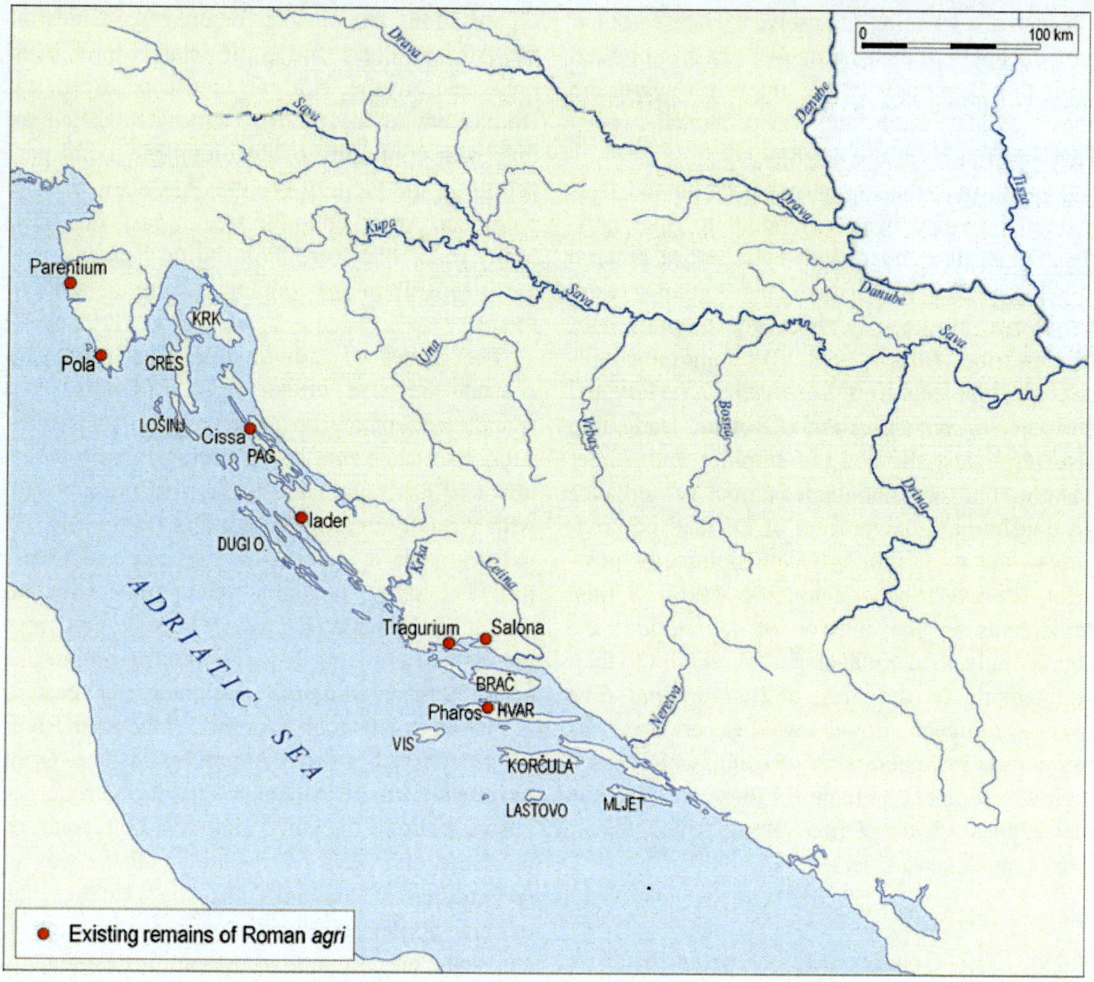

Fig. 3.3 Locations of remains of Roman *agri* on the eastern Adriatic coast (*Source* Regan 2003)

western part, primarily around *Pola* and *Parentium*. The economic basis was agriculture on parcels 20 ha or larger (Matijašić 1988). The spatially largest coastal province was *Dalmatia*, which was also organised in the first century. The province united the coast with the entirety of its Dinaric mountainous hinterland. During the era of Roman rule, the borders of the province didn't change, aside from the broad hinterland area north of the Krka River being sliced off and made into the province of *Liburnia*. The borders were a reflection of the political need to surveil and conquer the Illyrians, against whom the Romans often acted militarily.

Two legions of Roman soldiers stationed in Dalmatia are evidence of this. *Burnum* in Bukovica (in the hinterland of Zadar) was the headquarters of the 11th legion, and *Tilurium* (Trilj in the central Dalmatian Hinterland) was the home of the 7th Legion (Matijašić 2009). After a successfully undertaken pacification, near the end of the first century, the legions left *Dalmatia*. Afterwards, security, primarily of road traffic, was maintained by auxiliary military units called cohorts until the third century, when the system came under the guard of *Beneficiarii* security forces (Zaninović 1997). Illyrian tribes stayed true to their transhumant pastoral roots as a means of sustaining themselves. Their seasonal movements with their herds were best expressed from the Dinarides in the interior toward the coast (to the south and west) and the peri-Pannonian area (in the north).

Due to this, the northern border of the Dalmatian province was extended to the peri-Pannonian area and included the entire Dinaric hinterland. The main roads in *Dalmatia* were transverse, in keeping with transhumance of Illyrian tribes (Rogić 1974). By stationing military encampments from *Burnum* to *Tilurium* and the fort *Bigesta* (Ljubuški), Roman leadership also physically divided the summer and winter pasture areas and maintained direct surveillance on transhumant movements of Delmati communities—the most militarily and politically powerful Illyrian tribe[6] (Zaninović 2007). Urban settlements in the interior of *Dalmatia* were significantly weaker developed in relation to their counterparts on the coast or in *Pannonia*. The most significant growth was experienced by settlements in contemporary Bosnia, which based their development on mines or thermal sources in the valleys of the Drina, Sana, Una, Vrbas, Bosna, and Lašva rivers.

3.2.4 The Geographical Heritage of Roman Antiquity—Cultural Landscape, Toponymy, and Regional Concepts

Individual traces of Roman organisations in the cultural landscape remain visible to the time of writing, primarily on the coast, while much of their legacy in the Pannonian belt was destroyed in late Antiquity by repeated "barbarian" incursions (which started in the second half of the fourth century). Much was simply lost to neglect due to the disintegration of the Roman organisations responsible for upkeep.[7] Cities on the Adriatic coast and some islands have been preserved in the cultural landscape from the Roman era to the time of writing, boasting an unbroken continuity of development. Land parcelling in the vicinity of cities has also, in large part, been preserved in the area of *agri* (Fig. 3.3). Many roads that were built during Roman times are also still in use (albeit in a more modern form).

The names of individual regions, primarily coastal, are also drawn from Antiquity. The spatial concept of Istria (as a region and peninsula) also came into being during Roman antiquity and has continued as such to present day, with virtually unchanged borders (to the edge of the peninsula at the mountains Učka and Ćićarija). The name of the contemporary Croatian region of Dalmatia (*Dalmacija)* also comes from the Roman era, and it is important to emphasise how Roman *Dalmatia* included the eastern Adriatic coast from the conflux of the Raša River (in eastern Istria) to the conflux of the Drim River (in northern Albania). The northern border passed through the peri-Pannonian belt, south of the Kupa and Sava river valleys, all the way to the conflux of the Sava and the Danube. The eastern border extended far into the interior, across the eastern parts of present-day Serbia and Kosovo, such that *Dalmatia* united the Mediterranean belt with the Dinarides in the interior. By the end of the third century, *Dalmatia* had shrunk to the right bank of the Drina and the Bay of Kotor.

From the fifth century onward, the concept of *Dalmatia* was additionally narrowed with the separation of the northern coastal belt, from the conflux of the Raša River (eastern Istria) to the northern edge of Ravni Kotari, today in the hinterland of Zadar (including the mountainous hinterland) into its own province. Such were the borders of *Dalmatia* when Slavic settlers began to arrive on the eastern Adriatic coast. The Roman concept of *Pannonia* as a region,

[6]The aforementioned military encampments were arranged in a line which extended across the transhumant paths of Illyrian herders (primarily Delmati), and enabled easy surveillance of their movements.

[7]Hydro-regulatory complexes and roads are typically the first things to fall into disrepair due to lack of maintenance.

however, didn't survive. Namely, with the coming of new groups of people to settle on the Pannonian Basin (Alans, Huns, and Ostrogoths), around the turn of the fourth century to the fifth century, an unsuccessful policy of Christianising and civilising ended up reflecting negatively on the Roman cultural landscape. Roman efforts to stabilise the settlement of new nomadic peoples did not yield significant results (Gračanin 2011).

References

Dugački V, Regan K (eds) (2018) Hrvatski povijesni atlas. Leksikografski zavod Miroslav Krleža, Zagreb

Džino D, Domić Kunić A (2013) Rimski ratovi u Iliriku—povijesni antinarativ. Školska knjiga, Zagreb

Gračanin H (2011) Južna Panonija u kasnoj antici i ranom srednjovjekovlju (od konca 4. do konca 11. stoljeća). Plejada, Zagreb

Gračanin H (2012) Zagrebačko područje do osnutka Biskupije. In: Goldstein I, Goldstein S (eds), Povijest grada Zagreba—knjiga 1: od prethistorije do 1918. Novi liber, Zagreb, pp 2–29

Kuntić-Makvić B (1997) Grčka i rimska starina. In: Supičić I (ed) Hrvatska i Europa – kultura, znanost i umjetnost: svezak I. Srednji vijek (VII – XI. stoljeće) rano doba hrvatske kulture. AGM, Zagreb, pp 73–91

Matijašić R (1988) Ageri antičkih kolonija Pola i Parentium i njihova naseljenost od I. do III. stoljeća. VPA, Zagreb

Matijašić R (2009) Povijest hrvatskih zemalja u antici do cara Dioklecijana. Leykam international, Zagreb

Novak G (1944) Prošlost Dalmacije, knjiga prva: od najstarijih vremena do Kandijskoga rata. Hrvatski izdavalački bibliografski zavod, Zagreb

Regan K (ed) (2003) Hrvatski povijesni atlas. Leksikografski zavod Miroslav Krleža, Zagreb

Rogić V (1965) Položaj Senja i gravitacija: historijsko-geografski i suvremeni odnosi. In: Glavičić M (ed) Senjski zbornik, vol 1. Gradski muzej Senj—Senjsko muzejsko društvo, Senj, pp 7–21

Rogić V (1969) Rapska Otočna Skupina. Geogr Glas 31:109–124

Rogić V (1974) Historijsko-geografski uvod. In: Cvitanović A (ed) Geografija SR Hrvatske—knjiga VI: Južno hrvatsko primorje. Školska knjiga, Zagreb, pp 5–9

Rogić V (1977) Geografski aspekt podunavsko-jadranskih razvojnih osovina jugoslavenskog prostora u retrospektu i prospektu. In: Vasović M (ed), Zbornik X. jubilarnog kongresa geografa Jugoslavije, Sept 1976. Savez geografskih društava Jugoslavije—Srpsko geografsko društvo, Beograd, pp 252–258

Rogić V (1990) Regionalna geografija Jugoslavije knjiga 1—prirodna osnova i historijska geografija. Školska knjiga, Zagreb

Suić M (1976) Antički grad na istočnom Jadranu. Sveučilišna naklada Liber, Zagreb

Suić M (1997) Kontinuitet urbane kulture. In: Supičić I (ed) Hrvatska i Europa – kultura, znanost i umjetnost: svezak I. Srednji vijek (VII – XI. stoljeće) rano doba hrvatske kulture. AGM, Zagreb, pp 73–115

Zaninović M (1997) Iliri u antici na hrvatskom tlu. In: Supičić I (ed) Hrvatska i Europa—kultura, znanost i umjetnost: svezak I. Srednji vijek (VII – XI. stoljeće) rano doba hrvatske kulture. AGM, Zagreb, pp 55–71

Zaninović M (2007) Ilirsko pleme Delmati. Ogranak Matice hrvatske Šibenik—Gradska knjižnica „Juraj Šižgorić" Šibenik, Šibenik

4 Early Middle Ages and the Croatian State—Spatial Organisation and a New Cultural Landscape

Abstract

The transition from Antiquity to the Middle Ages was marked by deep-rooted changes throughout Croatian territory as well as the surrounding area, which were reflected in all segments of society—political, economic, and cultural. The nearly insurmountable differences between Roman civilisation and the newly arrived peoples led to a break in the developmental continuity of the cultural landscape. This overall change is reflected in decline and deurbanisation. This was mostly expressed in *Pannonia*, which was the area that was most exposed to invasion by migratory peoples arriving in Europe from the East. Conversely, the majority of littoral *Dalmatia* was spared from invasion, which mainly allowed the continuity of the urban cultural landscape to continue. This landscape change expressed the duality of Adriatic and continental Croatia, due to both environmental and political conditions, and organisation. The continuity of the development of coastal cities and environmental advantages, such as contact between Mediterranean lowlands and high mountain pastures (that enabled transhumance), were key elements that led to the birth of the first political core of the Croatian state in the area of northern Dalmatia. This is confirmed by the toponym *Hrvati,* the great density of *županas*—the first units of territorial organisation—and the seat of the first Croatian princes in the vicinity of Split. The area to the north of Gvozd, called Slovinje (Slavonia), was predominantly politically weak and unorganised in the Early Middle Ages. This can be seen in the lack of a basic territorial unit, i.e. župa, and churches.

Keywords

Croatia · Early Middle Ages · Župa · Geopolitical core · Dalmatia · Slovinje · Slavonia

B. Fuerst-Bjeliš and N. Glamuzina, *The Historical Geography of Croatia*, Historical Geography and Geosciences, https://doi.org/10.1007/978-3-030-68433-4_4

The Church of the Holy Cross. Nin, northern Dalmatia The pre-Romanesque Church of the Holy Cross was built in the ninth century. It has remained untouched since its construction and represents the most valuable preserved monument of early Croatian/Christian architecture. During the reign of Croatian kings, it served as a court chapel for the nearby duke's court (Photograph: Siniša Sakoman. Reproduced with permission)

The transition from Antiquity to the Middle Ages was marked by deep-rooted changes of Croatia as well as the surrounding area, which were reflected in all segments of society—political, economic, and cultural. Nearly insurmountable differences between Roman civilisation and newly arrived peoples led to a break in developmental continuity of the cultural landscape. During the entire Early Middle Ages, no newly founded European state (Croatia included) would reach the level of development of the Roman Empire.

4.1 Deurbanisation and Decline of the Roman Cultural Landscape

The deterioration of civilisations that dominated during Antiquity was the main cause of the deterioration of the basic elements of the Roman cultural landscape. The internal crisis within Roman society grew in strength throughout the last centuries of Antiquity. It manifested in the internal degeneration of the socio-political and economic systems. Obvious examples of this were the privatisation of public *agri* and the ability to buy political positions (on the local, regional, and even state levels). Attempts to regulate relations by implementing the Colonate system, tax reform, recognition of equal rights for the Christian Church, and military and administrative reforms were not successful. The final blow came in the form of foreign military invasion, in particular by the Huns (fifth century), and plague epidemics (foremost in 542 AD).

These two events hit hardest in *Pannonia*, which was the area that was most exposed to invasion by migratory peoples arriving in Europe from the East. Nearly all the elements of the cultural landscape (including cities themselves) were totally erased. The main consequence of this was an expressed regressive trend in the population that additionally affected the disappearance of cities (primarily in the Pannonian Basin), which had been the most important elements of Roman society and the cultural landscape. Political insecurity, already present in *Pannonia* in the second century, spurred the process of fortification of the surviving settlements, and even of large *villae* belonging to nobles or to the Empire itself. During the transition from the fifth to the sixth century, following the fall of the Western Roman Empire, the Eastern Roman (Byzantine) Empire invested a great amount of resources and efforts to comprehensively build new fortifications and renew and update existing ones (e.g. forts, fortresses, watchtowers, and guard towers), and repair existing city walls and build new ones where necessary (Gračanin 2011).

One of the attempts to solve the crisis and stop the regression, at least in terms of population and economics, was to allow peaceful colonisation and planned settlement of "barbarian" peoples in the parts of the Empire which had been hardest hit by the crisis. This could aid in solving the problem of a lacking workforce and population eligible for military service. Such policy, undertaken from the fourth to seventh centuries mainly in the Pannonian part of contemporary Croatia, had even drastic consequences for the cultural landscape. Namely, conflicts among the newly settled peoples arose, which, along with rebellions against Roman rule, had an exceptionally destructive influence on certain elements of the cultural landscape (Rogić 1990).

Conversely, the majority of *Dalmatia* was spared from "barbaric" invasion, which allowed continuity in the cultural landscape. Due to the bridge role of *Dalmatia*—linking the Eastern Roman Empire to the Apennine Peninsula—it was vigorously defended, further ensuring continuity of the cultural landscape. The arrival of refugees from *Pannonia* in *Dalmatia* and expressed migration of the population to islands and fortified coastal cities were linked with the efforts to strengthen the fortification systems of Dalmatian cities. The increasing sense of insecurity demanded stronger development of agriculture in the immediate hinterland of the main coastal urban centres (*Salona*, *Narona*, *Iader*, and *Epidaurum*), which had an important role as trade centres, harbours, and as production centres due to the development of textile workshops (Posavec 2007). The policy of stabilisation was dealt a deathblow when the Byzantine border on the Danube was breached by the arrival and invasion of the Avar-Slavic tribal alliance, who flooded into the Roman provinces on the eastern shore of the Adriatic during the sixth and seventh centuries (Rogić 1990).

4.2 Historical–Geographical Implications of Slavic Colonisation

Slavic colonisation, as we know from the oldest Byzantine sources, took the form of peaceful colonist incursions—at first. Avar-Slavic invasion campaigns soon followed, which further spread the settlement of new colonists into Roman provinces. Slavic peoples, like the other peoples who were new in the area, did not have advanced farming techniques or a developed urban culture. Their arrival also marked a return to the domination of primitive rural agriculture.[1] Slavs depended on mobile methods of animal husbandry and agriculture; thus, they had a constant need to find new agricultural areas. Conversely, the Avars depended on large numbers of subject farmers, who played the role of infantry during wartime, as opposed to the Avar warriors themselves who were primarily horsemen. The Avars could probably assemble the strongest cavalry force in the European continent in their heyday. This assertion is supported by the great number of Avar horseman graves that have been found (Sokol 1997).

In the forging of the Avar-Slavic alliance, certain key social differences, between the noble herder horsemen class and the subject farmer footmen class became clearly defined. The domination and power of the noble horsemen class resulted in the emergence of a hereditary aristocracy which would have a spatial reflection in the way that they organised territory under their control. The complex and long-term territorialisation of this hereditary societal organisation would, in its final form, make the basis of the first *župa* (the forebear of the modern Croatian territorial unit *županija*), which were led by a local tribal leader called a *župan*. The fundamental role played by hereditary nobility in the first forms of Slavic societal organisation is shown by the fact that the concept of župa was derived from the position of župan, and not the other way around (Rogić 1990).

To date, archaeological finds indicate the large concentration of Avars in the Pannonian Basin, primarily in Srijem and Baranja in the east, and in the upper part of contemporary Croatia along the Drava River in the west. Moreover, the numerical dominance of Slavs in the Avar Khanate has been confirmed by analyses of Avar-Slavic graves in the Pannonian area. These analyses indicate the small share of members of proto-Avar groups in the total population: only 6–10%. As the Avars transitioned to a sedentary lifestyle, the aforementioned cultural differences between them and the Slavs diminished, providing the basis for their ultimate slavicisation. Slavicisation spread from the core of the Pannonian area throughout the Avar Khanate and beyond, and the process was remarkably successful among autochthonous land-cultivating peoples with lifestyles similar to Slavs (Rogić 1990; Gračanin 2011).

4.3 Primary Geopolitical Nuclei and Territorial Organisation in the Context of New Environmental Values

The political core of the first Croatian state began to form during the Early Middle Ages. The meagre historical writings left behind by Byzantine chroniclers were based on sources from eastern Adriatic coastal cities. The small number of documents, the fact that they were rounded up nearly three centuries later, and the way they were written from the perspective of Romance peoples, calls their objectivity into

[1]The primarily agricultural character of Slavs shows in the fact that Slavic agricultural terminology can be found in Albanian, Romanian, and even Greek languages.

question. Based on these documents, despite their small number and questionable objectivity, we can come to some conclusions about the first Slavic state communities. Chroniclers clearly distinguished the idea *Sclavinae*[2] from the better and more firmly defined territories such as Croatia, Pagania, Zahumlje (Zachlumia or Hum), Travunia, Duklja (Dioclea)/Zeta,[3] and Raška (Rashka).

The political core of the first Slavic states on the eastern Adriatic coast encompassed two main, geographically different areas, i.e. two different types of natural environment. The areas were dominated by the main type of economy of Slavic peoples, primarily based on transhumance. Farming played a smaller role. Historical documents from the end of the Early Medieval Period (eleventh century) speak of underdeveloped agriculture in the coastal belt (Klaić 1990). Transhumance[4] demands complementary winter and summer pastures, therefore, it makes sense that the political cores of the first Slavic states encompassed two different environments: the Mediterranean as an area for winter pastures, and the Dinarides as summer pastures. This environmental duality was essential for maintaining and developing transhumance, and the newly formed Croatian state was a prime example thereof (Fig. 4.1).

First, the Mediterranean area of northern and central Dalmatia was a winter pasture area. The fact that these pastures were found in the hinterland of coastal Roman cities also dictated the location of the seats of the first Croatian leaders. This is confirmed by the toponym *Hrvati* in Ravni Kotari near the city of Nin, as well as the seat of the first Croatian princes in the vicinity of Split.

Second, the Dinarides in the interior had the role of summer pastures, whereby three mountain *župas* (Lika, Gacka, and Krbava) played key roles. Their economic and strategic importance also led to the emergence of a new type of powerful leader—*ban*[5]—whose task was to oversee these important summer pasture areas.

The bearers of the pastoral type of economy were members of the hereditary tribal aristocracy, from whose ranks would come the first leaders (bans, princes, and kings). The aristocracy initiated and supported the first territorial organisation in the areas settled by Slavs. This meant, apart from the formation of a state, the application of a new form of the organisation—župa—as the new primary territorial unit of the Croatian state. Namely, the earliest historical accounts, from the eighth, ninth, and tenth centuries, mention the role of the župan as a leader,[6] while župa, as an area administered by a župan, is first mentioned in later documents (typically after the eleventh century). Accordingly, it is likely that the župan represented the privileged land-owning class of the population who had certain privileges in their župa, such as grazing rights (Fuerst-Bjeliš 1996).

The selection of župans indicates the process of class division of Croatian society, by which the hereditary aristocracy also became military leaders (Nazor 2005). The power of the pastoral–tribal structure is reflected in its firm organisation, high mobility, and flexibility. At the same time, the indisputable authority of the warrior class was solidified, allowing them to impose their leadership upon the agricultural Slavic population and the remnants of autochthonous Illyrian communities without difficulty. The subdued sections of the Slavic population began to identify with the

[2]Slavic tribal federations. They became the nuclei of the future south Slavic states.

[3]From the twelfth century on, the name Zeta prevailed.

[4]Transhumance is understood to include seasonal movements with livestock in search of areas with good pastures: toward the warm Mediterranean areas in the winter; and into the high Dinaric mountain pastures in the summer. In the Dinarides, the most full and lush development of grass and other pasture vegetation takes place during the summer, which is typically a dry period on the coast.

[5]The ban was the highest ranked noble in Medieval Croatia, Slavonia, Bosnia, and in border areas of Hungary. The title "ban" probably comes from the Turko-Mongolic word *bajan*, meaning rich or landed, and probably entered the Croatian language due to interactions with the Avars. The ban was appointed by the king as his direct representative, sometimes with co-ruler status. The title roughly equates to duke.

[6]Župan is a lower level title. They were formally named by the ban.

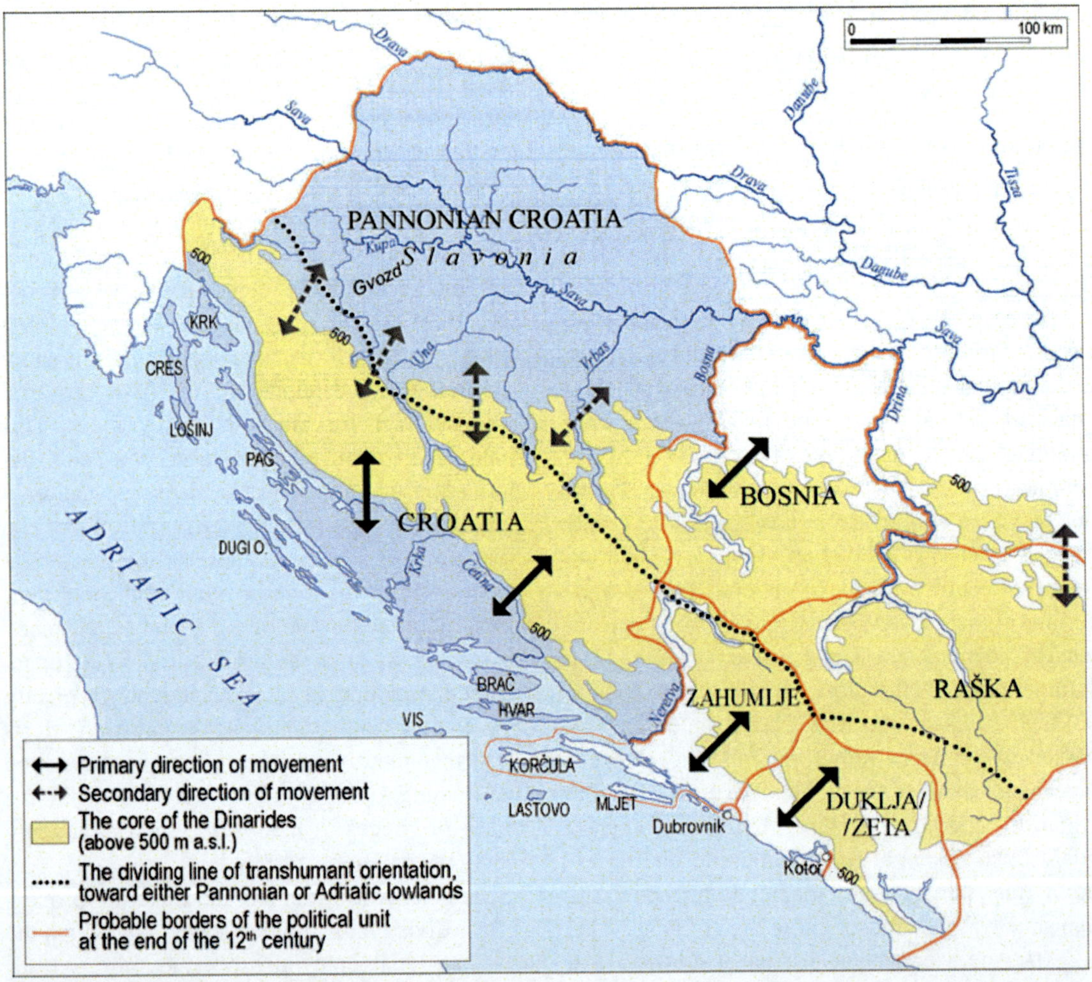

Fig. 4.1 The importance of environmental duality in terms of transhumance that linked the first Slavic territories together, and the formation of the first Croatian geopolitical core (after Rogić 1982)

warrior class that ruled them, which was reflected in the acceptance of common names and assimilation. Similar patterns are found in the ethnogenesis of Croats, Bulgars, and Serbs.

Although Slavic peoples also practiced farming, it had a secondary role in their economic structure. Farming agriculture was based on the cultivation of basic grains in the area of Ravni Kotari in the hinterland of Zadar, the Dalmatian flysch zones, Vinodol in Kvarner Bay, and on the islands. Products of animal husbandry (meat, leather, and tallow) were of higher value and were used to trade for agricultural products, which came to Roman coastal cities from the southern parts of the Apennine Peninsula. Additionally, trade between Croatians and Roman coastal cities was a necessity due to weak local productivity and the need for food. This is one of the main reasons that the coastal cities of Antiquity succeeded in preserving developmental continuity, as opposed to the large Pannonian centres that were not rebuilt after falling to invaders. During the period of Slavic colonisation, coastal cities retained maritime trade relations with cities on the Apennine Peninsula, through which they maintained relations with other parts of the Mediterranean (Rogić 1982, 1990, 1994).

4.4 Beyond the Nucleus—Conceptualisation of Croatian Historical Regions

In the Early Middle Ages the first borders of the Croatian state began to take shape. It extended from the eastern part of the Istrian Peninsula to Učka in the swampy zone of the Raša and Boljunšćica rivers. It is in this area that the border zone *Krajina* was formed. It later came under German feudal administration in the eleventh century and was referred to as the Duchy of Merania.[7] In the southeast, the border was in the area of Poljica in central Dalmatia, where the border with neighbouring Pagania (a principality along the Neretva River) ran along the lower part of the Cetina River. The border in the north toward Slavonia was connected to the geographical concept of Gvozd.

Medieval sources mention Gvozd as the classic concept of a border area with Slavonia, but this has yet to be completely clarified. Regardless of differing understandings of its exact location and scope, Gvozd marks the border area between the Dinarides and the peri-Pannonian area (see also Fig. 4.1). The meaning of the word *gvozd* in old Croatian tongue is an iron mountain (*Alpes ferrae*), which describes the natural environment of the border belt—relief, forestation, mines, etc. (Rogić 1990). The border area of Gvozd had dual significance.

First, transhumant paths passed through the area going from the Dinarides (in the warmer part of the year) toward the peri-Pannonian area (in the colder part of the year). Second, this is evidence of the continuing awareness of continuity with Antiquity because, during the era of Roman rule, the border between the provinces *Dalmatia* and *Pannonia* was drawn through this area (Kordun and Banija regions).

The area to the north of Gvozd, called Slovinje (Slavonia), was predominantly politically weak and unorganised in the Early Middle Ages. This can be seen in the lack of a basic territorial unit, i.e. župa, and churches. Archaeological findings show migrations from Croatia to Slavonia during the Early Middle Ages, primarily from the start of the tenth century (Gračanin 2011). Good conditions for the development of agriculture, especially for growing crops, were key attractive elements that strengthened the share of crop cultivation in the total economic structure. Stabilisation of settlements and economic opportunities gave an additional incentive to population growth. Additionally, the Croatian state, after converting to Christianity and coming under Frankish rule, was charged with spreading Christianity into neighbouring areas.

Frankish rule at the turn of the eighth century to the ninth century directed the Church organisation into the southern part of the Pannonian Basin. The role of the Croatian state in proselytising was outlined in historical documents which tell as to how the Croatian bishop came to lead the Diocese of Siscia (Sisak). The founding of the Diocese of Zagreb in the area of Slovinje at the end of the Early Middle Ages provides additional evidence of this. This event can be understood as the final step in the integration of Slovinje into the Croatian state organisation (Fig. 4.2). This probably took place during the tenth century, although to what extent the process actually ran its course by the end of the Early Middle Ages is debatable (Margetić 1977, 1997; Suić 1977).

The concept of Slovinje of that time encompassed the area of the far southwestern edge of the Pannonian plain, and stretched to the Drava River in the north, the Sutla and Kupa rivers in the west, and the outer edge of the Slavonian Mountains in the east. The southern boundary was Gvozd, stretching along the peri-Pannonian belt from the Kupa River to the Bosna Valley.

During the Early Middle Ages in the area known in Antiquity as the province of *Dalmatia*—which was then the largest province on the eastern Adriatic coast, boasting five Roman coastal cities (Kotor, Dubrovnik, Split, Trogir, and Zadar) and three island cities (Osor, Krk, and Rab)—came under Byzantine rule.

[7]The expression *Krajina* is, according to Rogić, the oldest written confirmation of the idea of a border area, to which the inscription on Baška tablet also attests, wherein Prince Kosmat is mentioned as leader of "the entire *Krajina*", i.e. the entire territory of contemporary Kvarner Bay.

Fig. 4.2 Integration of the Pannonian territory Slovinje/Slavonia (tenth century) and Pagania (eleventh century) into the Croatian state core (*Sources* Regan 2003; Dugački and Regan 2018)

Namely, the concept of Dalmatia was reduced to only the part under Byzantine control in the eastern Adriatic, underlying continuity from Antiquity and creating the basis for contemporary Dalmatia. Dalmatia, under Byzantine control in the Early Middle Ages, also included the Adriatic islands along with the eight aforementioned cities. Only the small islands of the Šibenik Archipelago along the coast and large parts of the large Kvarner islands belonged to Croatian territory during this period. The islands of Brač, Hvar, Korčula, and Mljet belonged to Pagania, and would only later (eleventh century) be integrated into Croatia. The interior of the Dinarides, which was settled by Slavs, completely lost continuity with Antiquity and the name Dalmatia fell out of use as a geographical term for the area.

Byzantine Dalmatia was not territorially continuous and encompassed a narrow coastal and island belt. It began to be regarded primarily as a maritime region. In keeping with Byzantine Dalmatia's shrunken area, it was also administratively divided into Upper (southeastern) and Lower (northwestern) Dalmatia. All cities of Byzantine Dalmatia enjoyed developmental continuity from Antiquity, apart from Dubrovnik—founded in the seventh century by Roman refugees from destroyed *Epidaurum*. Ethnic and cultural differences between Roman and Croatian inhabitants, as well as politics, did not allow for the integration of the former into the Croatian state. The Roman population was the bearer of urban culture, while the Croatian population was concentrated in rural settlements. During the Early Middle Ages, however, increasing economic integration reduced the aforementioned differences, little by little, and the process of ethno-social integration of the coastal cities with the hinterland began (Rogić 1990). The integration also had a political dimension, visible in that the Byzantine Empire temporarily gave over the administration of Dalmatian cities to Tomislav—the first Croatian king. The increasing influx of Slavs (Croats) into Roman cities is also indicated by individual documents from the tenth century, in which citizens, and even members of city leadership, with Slavic names are listed (Novak 1944).

The exception was the area of Istria, in which both Roman cities and rural settlements enjoyed continuity from Antiquity, together with new Croatian rural settlements. Here, economic relations were maintained between Roman cities and Roman villages (Suić 1976). The first mention of the spontaneous settlement of Slavs in Istria was noted at the turn of the sixth century to the seventh century and, at the turn of the seventh century to the eighth century, Istrian Slavs were incorporated into the Byzantine military structure. The Byzantine military organisation was concentrated primarily in the northern and central parts of the peninsula, and a dense network of fortifications was raised to defend against attacks from the north. Later, the Istrian area became part of the border of the Frankish Empire (788–952). Frankish administrators in Istria would also start the systematic settlement of Slavs into areas of unworked land, taken from Istrian communes. Although the influence of Venetian politics was on the rise in the first half of the tenth century, German rulers, for whom the area of Istria represented the most suitable point of access to the Mediterranean for the Austro-Germanic area, were of special importance.

In 952, the Bavarian Archduke Heinrich I took the Duchy of Friuli from the Italic king, which contained Istria, Verona, and Aquileia. Germanic influence would be established by Carinthian leaders from the Eppenstein family (starting in 995) as well as the direct influence of the German emperor. This enabled the transfer of the most advanced type of feudalism, which characterised Central Europe. Although Germanic influence did not spread, German craftsmen settling in cities gave an additional boost to development (Goldstein 1997). Istria would remain a part of the Holy Roman Empire, in which German royalty and aristocrats had the main role, until the end of the Early Middle Ages.

4.5 Župa (*Comitatus*)—The Primary Territorial Unit

The long-term process of territorialisation had its spatial reflection in the shaping of the župa[8] (Fig. 4.3). They were organised according to the principle of spatial logic, based on an independent, sustainable subsistence economy. The following cultural–ecological elements had a key role thereby: a balance between the characteristics and capacity of the natural environment; manner of exploitation; population variables; and ecological sustainability of such systems. A župa was a community, built along hereditary or noble lines, which grew into a territorial administrative unit via a long-term process of territorialisation. The transformation of župa from various forms of hereditary/noble organisations to basic territorial administrative units was closely linked to the transition from familial organisations to stronger forms of political organisation, which took place in the era of the Trpimirović dynasty.

A župa was typically named for the area in which it arose (according to nearby rivers, valleys, cities, etc.) and, as such, it is clear that the concept of župa was not just linked to the tribal community itself. Based on the size and characteristics of the area which they encompassed, it can be concluded that they made up a defined and independent territorial unit in terms of the subsistence economy, consisting of rural

[8]Referred to as *comitatus* in Latin historical sources.

Fig. 4.3 Župas (*comitatus*) in Croatia according to Constantine VII Porphyrogenitus (tenth century) (after Fuerst-Bjeliš 1996; adapted from Regan 2003)

municipalities. This would have been a fertile and protected spatial whole, bordered by natural (orographic or hydrographic) boundaries—located in a valley area of karst polje—and ideal for both animal husbandry and crop cultivation economies (Fig. 4.4). More than 70% of ancient Croatian župas were organised along karst poljes or river drainage basins, according to the spatial reality of a given area (Fuerst-Bjeliš 1996).

The first župas were direct political–administrative reflections of geographical reality and, as such, they became the basis of the territorial organisation of the Early Medieval Croatian state. The territory of a župa encompassed several villages, whereby the largest village had the role of a centre, toward which the remaining rural settlements gravitated. The territory of the original Croatian župas was incorporated into later forms of Croatian territorial organisation: ecclesiastic as well as administrative (Fuerst-Bjeliš 1996).

It is important to note how župas in Istria did not encompass large areas, because autochthonous political leadership was not shaped by prominent local leaders who surrounded themselves with župans—as was the case in the Early Medieval Croatian state. Namely, Frankish (and later German) administration was ushered in with the arrival of foreign (primarily Germanic) feudal lords, who prevented Slavic župans from

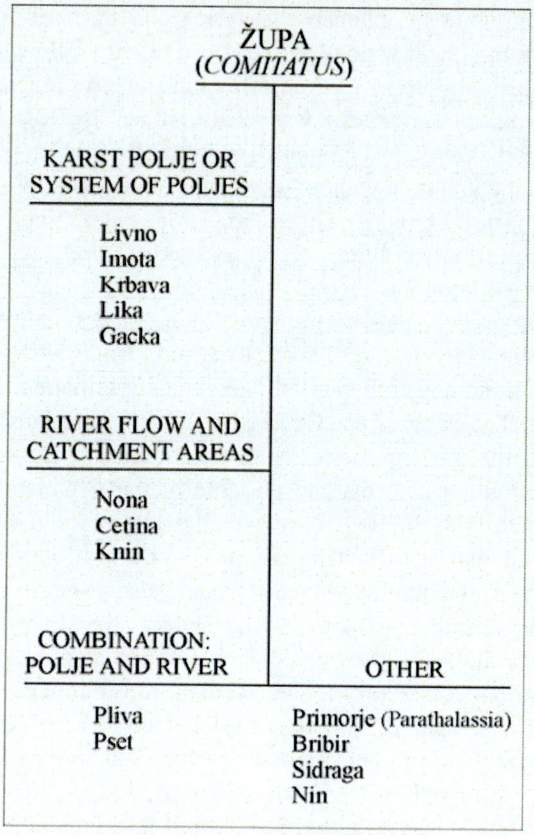

Fig. 4.4 The territory of županas (*comitatus*) in Croatia according to Constantine VII Porphyrogenitus, in relation to environmental elements (after Fuerst-Bjeliš 1996; adapted from Glamuzina and Fuerst-Bjeliš 2015)

becoming nobles in their own right. Accordingly, while župans in the Croatian state were climbing the social ladder by spreading their territory and establishing hereditary lines, the opposite was happening in Istria (Levak 2005).

4.6 Pagania—Integration of Bordering Principalities

During the transition from the Early to High Middle Ages, the territory of the Kingdom of Croatia spread following the integration of Pagania[9] into a united Croatian state. Pagania was one of the most important *Sclavinae* states, which extended throughout the area between the Cetina and Neretva rivers, including the large central and south Dalmatian islands (Brač, Hvar, Lastovo, Vis, Korčula, and Mljet). The territory of Pagania divided Upper and Lower Dalmatia, and because of this it had great strategic significance (see also Fig. 4.2).

Pagania lorded over shipping along the eastern Adriatic coast during the Early Middle Ages, and its pirates often attacked and plundered Venetian ships sailing in the area. In order to oversee maritime shipping, there was a network of fortified settlements at the foot of Biokovo Mountain, which is mentioned in records from the Early Middle Ages. The first two, Mokro and Verulja (Berulia—in the area of Makarska and Brela), were placed at the point where the continental landmass is closest to the island of Brač. The other two fortified settlements, Ostrok and Slavineca (in the area of Zaostrog and Drvenik), were raised where the continental landmass was closest to the island of Hvar, in order to observe shipping from the Korčula Channel toward the Neretva Delta.

The placement of these four settlements was in keeping with the manner of coastal maritime shipping characteristic of the day, which was active from sunrise to sunset. Local/friendly ships were able to take shelter at the fortified coastal settlements in the event of bad weather or attack. Along with strategic oversight of shipping lanes, these forts also controlled points of ingress into the interior of Dalmatia and Herzegovina (Goldstein 2005). Accordingly, Mokro and Verulja had the additional strategic significance due to their proximity to the Croatian state, with which Pagania had a direct land border. Although links between the two political entities also existed during the Early Middle Ages—the Paganian prince Domagoj made an official visit to the Croatian throne—total political integration did not take place until the end of the Early Middle Ages.

There are no documents that indicate economic links, so the eventual political integration is difficult to explain as a consequence of economic factors. The main factors are likely found in political conditions which indicate the Early Medieval practice of territorial expansion of politically strong states (in this case the Kingdom

[9] In documents from the Early Middle Ages, this principality is also called *Arenta* (*Narenta*).

of Croatia) into the territory of their weaker neighbours. Political integration was certainly eased by the battle against their mutual enemies, such as the Venetians, who tirelessly meddled on the eastern Adriatic coast (Raukar 1997).

Individual documents from the Early Middle Ages also indicate the strategic importance of Pagania, such as a declaration from 1050 by which the Paganian prince confirmed the gift of one church on the island of Biševo.[10] Accordingly, Pagania had control over islands in the southern Adriatic, which were of exceptional strategic value and gave them control over shipping between the greater Mediterranean and the northern Adriatic (Klaić 1990).

In geographical terms, this enabled the easiest surveillance over maritime shipping, as well as links to the western Adriatic coast. At the time of writing, however, the question of how strong of a link Pagania had with the Slavic settlement of Gargana in the southern Adriatic remains unanswered (Nikolić 2005).

Additionally, these fact that Croatia, as a vassal of the greatest European power of the day (the Frankish Empire) and a Papal ally, had taken up the responsibility of spreading Christianity cannot be ignored. It just so happened that the conversion of Pagania overlapped with it being integrated into the Croatian state, so we can safely assume that the two events were linked. By uniting, both Croatia and Pagania strengthened their collective strategic position on the eastern Adriatic coast, in the face of the growing power of Venice.

4.7 Landscape Change—The Duality of Adriatic and Continental Croatia

The new cultural landscape, which took shape in the area of the newly formed Croatian state, was characterised by sparse settlement (5–10 inhabitants/km^2). Low population density was supported by extensive pasturing and primarily shifting cultivation, based on the fire-fallow (slash-and-burn) method. This kind of primitive form of agriculture was characterised by low productivity, and supported a primarily rural settlement pattern without urban centres. The littoral (the direct hinterland of eastern Adriatic cities) had the primary political, economic, and cultural role in the fledgling Croatian state. Cities were, above all, trade centres where wood, slaves, and animal husbandry products were exported, while agricultural and crafted goods were mainly imported.

The zone of economic and cultural influence of the Roman coastal cities on the narrow hinterland, which marked the core of the Croatian state, was limited. It was in the littoral of the Croatian state that the Roman world contacted the Slavic familial–hereditary and warrior–herder organisation (Rogić 1990). During the Early Middle Ages, the immediate surroundings of the aforementioned cities enjoyed the most intense use of space. Namely, all of the coastal cities continued to maintain and develop their *agri*.

Although the area of said *agri* was significantly reduced, along with a drop in population which happened during the transition from Antiquity to the Early Middle Ages, they continued to exist. During the invasions of "barbarian" peoples, as well as the first phase of Slavic colonisation, *agri* were targets of vandalous assaults by the newcomers. Due to their importance, however, citizens of coastal cities tirelessly restored them. The stabilisation of political and security conditions in the Early Middle Ages enabled further development and improvement of *agri*. Furthermore, the Church kept the seats of its archdiocese and dioceses in urban settlements, as was the custom in many other parts of Europe, and was able to offer help with the reconstruction of urban networks.

The archdiocese had seats in three Dalmatian cities, Dubrovnik, Split, and Zadar, which made sense due to the Roman tradition of ruling from coastal cities. The seat of various dioceses were, in contrast, located in smaller urban centres. In Istria, the cities of Poreč and Pula had this role (which they took on in late Antiquity), and two new dioceses were founded in the sixth century,

[10] A small protruding island near the island Vis. The declaration is preserved in clerical archives on the Italian Tremiti island chain, in the vicinity of the Gargano Peninsula, the part of the Apennine Peninsula that protrudes the farthest into the Adriatic Sea.

with seats in Novigrad (on the western coast of Istria) and Pićan on the eastern coast near the Raša River. In Kvarner Bay, from the sixth century to the eighth century there were dioceses in the island cities Krk, Rab, and Osor (on the island of Cres), while in the twelfth century a new diocese was founded in Senj. The greater part of the mountainous area, in the immediate hinterland of Kvarner Bay, was under the auspices of a newly founded diocese that had its seat in Modruš.

In northern Dalmatia, the function of the diocese during the Early Middle Ages was carried out in Nin (ninth century), Skradin (tenth century), and from the eleventh century also in Knin and Biograd. Trogir and Makarska were seats of dioceses in central Dalmatia and Ston in southern Dalmatia (Šanjek 1997). The distribution of dioceses and the seats themselves were closely linked with the distribution of population, and the level of economic development. The founding of new dioceses in Kvarner Bay, the mountainous areas, as well as in the hinterland of Dalmatian cities, illustrates the strengthening of the economy and population in these areas—which was in accordance with stronger agriculture and the solidification of the feudal social system. During the Early Middle Ages, the area of the Pannonian Basin was characterised by the continued destruction of the Roman cultural landscape. The destruction of the Avar Khanate at the hands of the Franks at the end of the eighth century greatly contributed to this as well as Frankish military campaigns against Ljudevit Posavski at the beginning of the ninth century, Moravia at the end of the ninth century, and fierce invasions by the Hungarians[11] at the end of the ninth century (Gračanin 2011).

4.8 Population Change as a Consequence of Slavic Colonisation

The settlement of Slavs into the area of the former Roman province of *Dalmatia* was followed by the suppression of the non-Slavic (more or less Romanised) Illyrian inhabitants deep into the interior. The strength and fighting spirit of the Illyrian tribes had been largely broken by five centuries of Roman rule. In accordance with the Roman policy of destroying the internal structure of Illyrian society during and after the conquest of a given area, the leading land-owning families were then offered the chance to keep what they possessed and incorporate themselves into the Roman state organisation, under the condition that they acknowledge Roman leadership. The remnants of an Illyrian society that continued to resist were driven into inaccessible areas that were poor for the development of animal husbandry. This meant that they didn't have the strength to oppose the Slavs who arrived in the Early Middle Ages and subsequently organised the first Slavic states.

In the Early Middle Ages, the term Vlach[12] (*Vlah*) came to be used to describe these Romanised shepherd societal communities. Their identity was (and still is) also associated with backwardness. The partial slavicisation of Vlachs began during the Early Middle Ages, and was spurred by conversion to Christianity and the development of Slavic political leadership. Preserved documents from that era, as well as from the High Middle Ages, show how the process of slavicisation itself did not finish during the Early Middle Ages (Rogić 1976). The slow incorporation of warrior–herder elements into the newly founded Slavic states in the Early Middle Ages shows the significance of this way of life and the characteristic mentality of opposition to the leaders of the new state organisation.

Namely, the Vlach population was also based on warrior–herder social organisation, and its forerunners, largely from the ranks of the familial–tribal aristocracy, refused to accept the leadership of Croatian aristocrats. In contrast, the broad Slavic land-cultivating section of society was very quick to accept the leadership of the newly arrived familial–tribal warrior element. Therefore, Dinaric Vlachs, surrounded by a majority Slavic population, were made up of

[11]Frankish historical sources from that era describe the Pannonian area as a "wasteland".

[12]Also called Wallachians. In Venetian tradition, they were called *Morlacci* (Moro-Vlach).

only dispersed remnants that identified themselves with a separate socio-economic organisation and legal status. The slavicisation of the Vlachs would increase significantly, in step with the increasing intensity of conversion to Christianity. This is shown by the first documents in which they identified themselves as Croatian Vlachs of Latin tradition (Rogić 1958, 1982).

Furthermore, slavicisation was emphasised in Dalmatia. In the Early Middle Ages, the spatial concept of Dalmatia had shrunk to a disjointed area of Roman cities and their immediate surroundings. The entirety of the remaining area outside of these cities was Croatian. The assimilation of Roman citizens into the Croato-Slavic masses (the earliest confirmation thereof is from the eleventh century) resulted in the synonymous use of the words Croatia and Dalmatia for the same area (Rogić 1990, 2003).

References

Dugački V, Regan K (eds) (2018) Hrvatski povijesni atlas. Leksikografski zavod Miroslav Krleža, Zagreb

Fuerst-Bjeliš B (1996) Historijsko-geografski pogled na pitanje prvobitne teritorijalne organizacije Hrvatske. Acta Geographica Croatica 31:115–124

Glamuzina N, Fuerst-Bjeliš B (2015) Historijska geografija Hrvatske. Sveučilište u Splitu, Split

Goldstein I (1997) Između Bizanta, Jadrana i Srednje Europe. In: Supičić I (ed) Hrvatska i Europa – kultura, znanost i umjetnost: svezak I. Srednji vijek (VII–XI. stoljeće) rano doba hrvatske kulture. AGM, Zagreb, pp 169–178

Goldstein I (2005) Naseljeni gradovi Porfirogenetove Paganije (De administrando imperio 36/14-15). In: Budak N (ed) Raukarov zbornik – zbornik u čast Tomislava Raukara. FF Press, Zagreb, pp 39–53

Gračanin H (2011) Južna Panonija u kasnoj antici i ranom srednjovjekovlju (od konca 4. do konca 11. stoljeća). Plejada, Zagreb

Klaić N (1990) Povijest Hrvata u srednjem vijeku. Globus, Zagreb

Levak M (2005) Tragovi društvenog ustroja istarskih Slavena u ranom srednjem vijeku. In: Budak N (ed) Raukarov zbornik – zbornik u čast Tomislava Raukara. FF Press, Zagreb, pp 55–73

Margetić L (1977) Konstantin Porfirogenet i vrijeme dolaska Hrvata. Zbornik Odsjeka za povijesne znanosti Zavoda za povijesne i društvene znanosti JAZU 8:5–88

Margetić L (1997) Hrvatska država u doba narodnih vladara. In: Supičić I (ed) Hrvatska i Europa – kultura, znanost i umjetnost: svezak I. Srednji vijek (VII–XI. stoljeće) rano doba hrvatske kulture. AGM, Zagreb, pp 197–214

Nazor A (2005) O ustroju hrvatske vojske u ranome srednjem vijeku. In: Budak N (ed) Raukarov zbornik – zbornik u čast Tomislava Raukara. FF Press, Zagreb, pp 83–108

Nikolić Z (2005) Najstarija slavenska naselja u južnoj Italiji. In: Budak N (ed) Raukarov zbornik – zbornik u čast Tomislava Raukara. FF Press, Zagreb, pp 75–82

Novak G (1944) Prošlost Dalmacije, knjiga prva: od najstarijih vremena do Kandijskoga rata. Hrvatski izdavalački bibliografski zavod, Zagreb

Posavec V (2007) Dalmacija u vrijeme Marcelina i Julija Nepota. Književni krug, Split

Raukar T (1997) Prostor i društvo. In: Supičić I (ed) Hrvatska i Europa – kultura, znanost i umjetnost: svezak I. Srednji vijek (VII–XI. stoljeće) rano doba hrvatske kulture. AGM, Zagreb, pp 181–195

Regan K (ed) (2003) Hrvatski povijesni atlas. Leksikografski zavod Miroslav Krleža, Zagreb

Rogić V (1958) Velebitska primorska padina – prilog poznavanju evolucije krškog pejzaža (nastavak). Geografski glasnik 20:53–110

Rogić V (1976) Socio-geografski aspekt dinarskog krša, dinarskog kulturnog areala i dinarskog brdsko-planinskog prostora. Geografski glasnik 38(1):253–269

Rogić V (1982) Historijsko-geografska osnova socijalno-kulturne diferencijacije vojnokrajiškog prostora. Geografski glasnik 44(1):23–38

Rogić V (1990) Regionalna geografija Jugoslavije, knjiga 1 – prirodna osnova i historijska geografija. Školska knjiga, Zagreb

Rogić V (1994) Geopolitical retrospect of Croatia. In: Crkvenčić I, Klemenčić M, Feletar D (eds) Croatia—a new European state, Zagreb, 1993

Rogić V (2003) Sveti Juraj – mikrogeopolitički retrospekt i prospekt. In: Babić S (ed) Spomenica Svetog Jurja. Udruga za zaštitu i promicanje razvitka Svetog Jurja, Rijeka, pp 9–27

Sokol V (1997) Arheološka baština i zlatarstvo. In: Supičić I (ed) Hrvatska i Europa – kultura, znanost i umjetnost: svezak I. Srednji vijek (VII–XI. stoljeće) rano doba hrvatske kulture. AGM, Zagreb, pp 117–146

Suić M (1976) Antički grad na istočnom Jadranu. Sveučilišna naklada Liber, Zagreb

Suić M (1977) Ocjena radnje L. Margetića „Konstantin Porfirogenet i vrijeme dolaska Hrvata". Zbornik Historijskog instituta JAZU 8:89–100

Šanjek F (1997) Crkva i kršćanstvo. In: Supičić I (ed) Hrvatska i Europa – kultura, znanost i umjetnost: svezak I. Srednji vijek (VII–XI. stoljeće) rano doba hrvatske kulture. AGM, Zagreb, pp 217–236

5 High Middle Ages—Economic and Political Stabilisation, Re-urbanisation, and Regional Differentiation

Abstract

The end of the Early Middle Ages in Croatia overlaps with the end of the era of the Croatian Trpimirović dynasty and the beginning of the union with the Kingdom of Hungary. The founding of the Hungarian–Croatian Kingdom—a new political entity in Europe—was linked with the creation of new political nuclei in continental Croatia and Slavonia, encompassing the area between Gvozd and the Drava River. Medieval Slavonia also had great strategic significance because it linked the area of Hungary to Adriatic harbours, thereby increasing trade opportunities. The expressed process of strengthening of the autonomy of Dalmatian Roman cities resulted in their exclusion from the Hungarian–Croatian state, with which they had been economically integrated. This is also a period of re-urbanisation and foundation of a network of royal fortified cities and royal županijas (counties). The chapter will also focus on changing concepts and definitions of Croatian historical regions, as well as the rise of the Republic of Venice, which would become one of the most powerful forces in the Mediterranean during the High Middle Ages, and the initial formation of the territory that would become the Republic of Ragusa (Dubrovnik). During the High Middle Ages, individual parts of contemporary Croatian territory were, for shorter or longer stretches, included in the territories of neighbouring states. The chapter draws attention particularly to interactions between Croatia and Bosnia in this period.

Keywords

High Middle Ages · Hungarian–Croatian Kingdom · Slavonia · Republic of Venice · Dubrovnik · Bosnia

B. Fuerst-Bjeliš and N. Glamuzina, *The Historical Geography of Croatia*,
Historical Geography and Geosciences, https://doi.org/10.1007/978-3-030-68433-4_5

Šibenik: the first street of the oldest Croatian city and Pellegrini-Tambača palace in detail, Dalmatia
Šibenik was founded in the ninth century and is the oldest city that was founded by Croatians. The Pellegrini-Tambača palace, partly built on the town's fifteenth century defensive walls, lies along the city's oldest street (from the thirteenth century). It is interesting among Croatia's palaces due to its combination of Renaissance and Baroque elements (fifteenth–eighteenth centuries). (Photograph: Borna Fuerst-Bjeliš. Reproduced with permission)

The end of the Early Middle Ages in Croatia overlaps with the end of the era of the Croatian Trpimirović dynasty and the beginning of the union with the neighbouring Kingdom of Hungary. The feudal socio-political system strengthened during the High Middle Ages, which led to feudal fracturing and eventually anarchy. Out of these processes of disintegration, three political wholes with different developmental tendencies emerged: the Hungarian–Croatian Kingdom; Dalmatian communes; and the interior of Istria.

The expressed process of strengthening of the autonomy of Dalmatian Roman cities resulted in their exclusion from the Hungarian–Croatian state, with which they had been economically integrated. The area of Gvozd became a border zone between the territory of Croatia and Slavonia on one side, which was under the influence of the Hungarian crown, and the area of Adriatic Croatia to the south of Gvozd on the other side, which was ruled by Croatian noble lines. The strongest invasion of Germanic political elements (especially that of the Habsburgs later in the era) into the area of Croatia during the High Middle Ages took place in the area of Istria.

The High Middle Ages were also an era of increasing political pressure on the edges of Croatian territory. This was due to both the political ambitions of Venice in the Adriatic and the rapid territorial expansion of the Banate of Bosnia (later the Kingdom of Bosnia), which came at the expense of Croatian territory in the Dinaric and Adriatic areas. The Bosnian state would, in this way, carve a piece away from the Early Medieval Croatian State—a process that would continue in the fifteenth century under the Ottomans (Rogić 1990; Raukar 2000).

5.1 New Geopolitical Nuclei and Development of the Geographical Concepts of Croatian Regions

The founding of the Hungarian–Croatian Kingdom (Fig. 5.1), as a new political entity in Europe, was linked with the creation of new political nuclei in Croatia. The area of Slavonia, which in that era encompassed the area between Gvozd and the Drava River (including northern and northwestern Bosnia and the eastern regions of Srijem and Baranja), became the main area of political focus in Croatian territory. The importance of the area can be seen in the role of the area's bans and dukes, who were often members of the Hungarian royal family.[1] The causes of moving the political core from the area of central and northern Dalmatia to the Pannonian Basin lay with political and economic factors.

Of the political factors, the most important was the union of Croatia and Hungary, a Central European state first led by the Árpád dynasty, then the Capetian House of Anjou, Sigismund of Luxembourg, and finally the Jagiellonian dynasty. The great distance of northern and central Dalmatia from the Hungarian political core and the close vicinity of Medieval Slavonia both supported the relocation of the political centre of Croatia into the Pannonian Basin. An additional reason was the presence of strong Croatian noble families in the area south of Gvozd, who might have grown to become political threats under the conditions of increasing feudal anarchy on the European continent. The process of political and economic independence of feudal lords was strengthened after the royal family signed the Golden Bull in 1242, which freed noble families from paying taxes and enabled them to do as they wished with their holdings (Magocsi 2002).

The most important economic factors were the gradual strengthening of crop cultivation in the economic structure. This process was also present in other parts of Europe, thanks to the application of different innovations[2] spread by monastic orders—especially the Knights Hospitaller, Knights Templar, and the Order of Cistercians—throughout their holdings. The spread of a network of monasteries belonging to the aforementioned monastic orders added an

[1]Typically, the crown prince was given this title and thus prepared for his future role as King.

[2]The most important innovation was, without doubt, the advent of two- and three-field systems of crop rotation.

Fig. 5.1 Croatia within the Hungarian–Croatian Kingdom (1102) (adapted from Regan 2003)

important economic and cultural dimension to the areas in which they were built, as well as helping to spread and maintain the faith. Medieval Slavonia also had great strategic significance because it linked the area of Hungary to Adriatic harbours, thereby increasing trade opportunities.

Geographical traces of the aforementioned political changes can be seen in the creation of royal županija (counties) with fortified cities as centres (Fig. 5.2). They most likely came about by reshaping and combining some smaller župas together during the High Middle Ages to make a larger županija. This meant that the župas themselves took over the role of smallest territorial unit (*kotar* or *distrikt*), land-owning unit (free noble municipality), and ecclesiastical territorial organisation unit (archdiocese) (Fuerst-Bjeliš 1996). The revival of local and regional transit between larger urban settlements, along communication routes from Roman times, is another indication of the population and economic development of this part of Croatia.

The geographical concept of Medieval Slavonia in the High Middle Ages was largely the same as it was in the Early Middle Ages. Although, in the High Middle Ages, the concept gradually came to be used also for the area east of Požega Valley, which had meanwhile been more firmly integrated into the Hungarian–

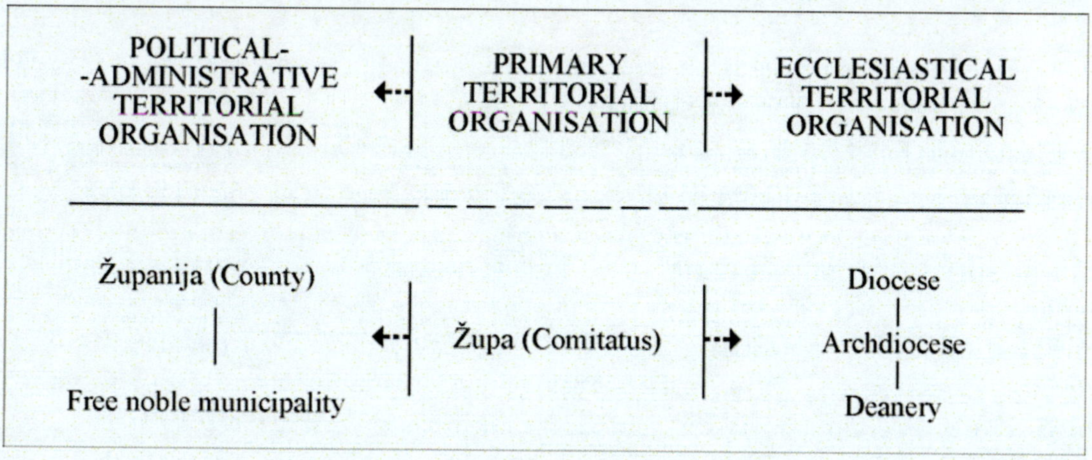

Fig. 5.2 Schematic display of the spatial continuity of territorial organisation during the Middle Ages (after Fuerst-Bjeliš 1996; adapted from Glamuzina and Fuerst-Bjeliš 2015)

Croatian state. Additionally, Slavonia was politically separated from Croatia and Dalmatia.[3] Medieval Slavonia was divided into two smaller parts (Upper and Lower Slavonia), which were linked by the transitory area of the Požega Valley. The area of Medieval Upper Slavonia encompassed old Croatian Slovinje in the west and was divided into eight županijas (Fig. 5.3); the seats of the županijas were in Zagreb, Varaždin, Križevci, and Virovitica. These four centres were also the main incubators of settlement and traffic hubs.

The remaining four županijas—Dubica, Vrbas, Sana, and Glaž—mainly covered the territory of contemporary peri-Pannonian Bosnia, and were of lesser significance. Upper Slavonia had better conditions for agriculture, and the main cultivation zones were in the foothills. These zones, thanks to their pedological and ecological diversity, offered the best conditions for the cultivation of traditional rural polycultures, which was the basis of agriculture for that era. Within the framework of the feudal economy, this was the main area of settlement (Rogić 1990). Individual parts of Upper Slavonia, such as the Međimurje Region, were already the subject of significant settlement density in this era[4] (Laci 1982).

Lower Slavonia, in the east, was an area where Slavic lands met Hungarian lands, and encompassed Baranja Županija (with Pécs as its main centre) and Vukovar Županija, encompassing the greater part of the Srijem Region. Late and poor use of the most valuable soils on the loess plains resulted in a lower developmental level of agriculture and the dominance of nomadic herding groups. The weaker development of crop cultivation was causally linked with the lagging development of permanent settlements in lowland loess areas. Due to this, settlements in the foothill belt, such as Pécs, developed into the main population centres. The second focal point of settlement development was the banks of the Danube, largely due to the river's important role in transit and trade. Feudal fortified towns were built to oversee river traffic, indicating the Danube's importance as a means of transit and transport.

[3]From 1260, in keeping with a decision of the Hungarian–Croatian King, Slavonia would be administered by a single ban, and Croatia and Dalmatia by another, effectively dividing the area of Croatia in political terms. Both bans would answer to the archduke (*herceg*), who held sway over the entirety of Croatia, Dalmatia, and Slavonia.

[4]This is confirmed by documents from the thirteenth century, which show a large number of ecclesiastical and feudal holdings, smaller settlements (mercantile), as well as the founding of the first cities into which craftsmen were settled according to plans.

Fig. 5.3 Organisation of županijas in the High Middle Ages (thirteenth century) (*Sources* Regan 2003; Dugački and Regan 2018)

5.2 Slavonia—Urban and Transportation Networks as Indicators of Political and Economic Change

The development of cities, as new elements in the cultural landscape of Slavonia, was an indicator of more significant development of this part of Croatia. In the second half of the fifteenth century, the urban network became more dense and, apart from free royal cities, market towns appeared. Of the 120 cities and market towns (which were known to exist in Slavonia in this era), around 100 of them were newly built settlements (Kurelac 2000). In the western part of the Pannonian Basin, e.g. Varaždin Županija, the main urban settlements had the role of centres in their respective areas. The standard distance between cities was the measure of a single day's walk, although this distance differed in different areas (Budak 2007).

Significant construction of fortified towns began in the foothill zones of Požega Valley, Medvednica and Kalnik mountains, and the Hrvatsko Zagorje Region, indicating a strong economic base and the existence of complex socio-political organisations. These well-fortified settlements showed their worth in short order during the Mongol invasion (1241–1242), which led to royal support for the construction of further fortifications (Magocsi 2002). The majority of the new fortified towns, built through the end of the High Middle Ages, would become centres of urbanisation. The strongest development during the High Middle Ages was found in Zagreb, which was the seat of an important diocese that covered a large territory, including Međimurje and Slovenian Prekomurje. Zagreb also had a strong economic role in this era.

Medieval Zagreb was, in fact, two cities, i.e. the royal fortified town of Gradec (also called Grič) on one hill and Kaptol, the centre of the diocese, on the neighbouring hill. In the mid-fifteenth century, the combined population was between 4,000 and 6,000. Because of its size, the city heavily influenced the entire Slavonian area. In the central part of Slavonia, Požega played a key role and most likely boasted a population of over 1,000. Along with fortified towns and royal cities, which were the highest rank among settlements, other types of settlements (of lower levels rank) also enjoyed significant development. In this case, lower rank means smaller towns located in important transit areas, such as bridges or mountain passes (e.g. Varaždin, Virovitica, Vukovar, Osijek, and Modruš). Market towns, such as Pakrac, Čazma, and Ivanić Grad, were numerous and important in this period; however, only a few grew to true importance and size—Ilok on the Danube, for example, boasted around 2,000 inhabitants at the end of the fifteenth century.

Market towns significantly differed in the number of inhabitants, but a rough average was 50 to several hundred. The bearers of these first humble embryos of Medieval urbanisation in the area of Slavonia were craftsmen and merchants, who immigrated from different parts of Europe—primarily from Italian- and German-speaking areas. There were also Francophone settlers as well as Jews (Grgin 2012). Cities were like islands that differed from the surrounding rural space in political, economic, and ethnic terms, as was the case in other parts of the Pannonian Basin. The area's autarchic agriculture, however, did not produce a surplus that could spur stronger development of the market or cities (Rogić 1990).

The founding and development of free royal cities in the area of Slavonia (Fig. 5.4), via royal decree, were a result of resistance from the urban classes (merchants and craftsmen) against the pressures of feudalism. During the thirteenth and fourteenth centuries, the first such royal cities arose—Varaždin (1209), Perna (1225), Vukovar (1231), Virovitica (1234), Petrinja (1240), Samobor (1242), Gradec (Grič) (1242), Križevci (1252), Jastrebarsko (1257), Zelina (1328), Trg ispod Ozlja (1329), Krapina (1347), and Koprivnica (1356). The greatest number of free royal cities were located along the edges of the Pannonian Basin, in the following regions: Podunavlje (the area along the Danube River); Podravina (the area along the Drava River); Posavina (the area along the Sava River); Hrvatsko Zagorje; Prigorje; Pokuplje (the area along the Kupa River); Žumberak. In these areas, these towns oversaw agricultural land and economically and strategically important transit routes. They were typically 30–70 km distant from one another, which indicates the size of their respective gravitational areas (Mohorovičić 2000).

Other urban centres also developed in the same era, such as Čakovec, Krapina, Čazma, and Đurđevac in the western part of Slavonia, and Đakovo, Požega, Ilok, and Vukovar in the eastern part. In contrast to Adriatic cities, Slavonian cities were not able to gain autonomy from the nobility, and sought royal protection for this reason (Raukar 2000). The economic strengthening of feudal aristocrats in the area of Slavonia (Babonići, Celjski, Frankopani, Gorjanski, Iločki, Talovci, and Zrinski families) was reflected in their growing political power and frequent internecine conflicts. The emergence and development of free royal cities are extremely significant because these were cores of urbanisation in the Pannonian Basin. The majority of such cities would grow to become significant urban centres in the following centuries.

The new political and economic core was linked to Croatia to the south of Gvozd by a network of primitive paths. There were several important passes where fortifications and urban

Fig. 5.4 The founding of free royal cities in Croatian lands during the fourteenth century

settlements were located. Two main routes linked the Pannonian core with the northern Adriatic through the Croatian Corridor. One led from Zagreb through Dubovac (a fortified settlement where Karlovac would be later built) and Modruš toward the northern Adriatic harbour Senj. A second route led from Dubovac toward the northern Adriatic. This one, however, led to Rijeka via Brod on the Kupa and Skrad. International trade of goods coming from continental Croatia and Hungary took place at the Adriatic harbours Bakar, Bakarac, Kraljevica, Novi, and Senj, giving the connecting roads a special significance (Budak 2007). The second main route also went from Zagreb, through Petrinja, Topusko, and Cetin toward Bihać and eventually Knin (the most important transport node south of Gvozd). From Knin, the route split three ways: one leading toward the Dalmatian cities Nin and Zadar, the second toward Šibenik, and the third toward the pass of Klis and Split (Fig. 5.5).

Two roads led to Hungary north of Gvozd: one from Zagreb through Križevci and Koprivnica; the other from Pakrac through Daruvar, Virovitica, and Barcs. These routes were also guarded by forts and fortified cities. The primitive paths toward Austrian lands were of lesser importance, i.e. from Zagreb through Varaždin to Styria; and from Ozalj through Metlika, Novo Mesto, and Ljubljana to Trieste.

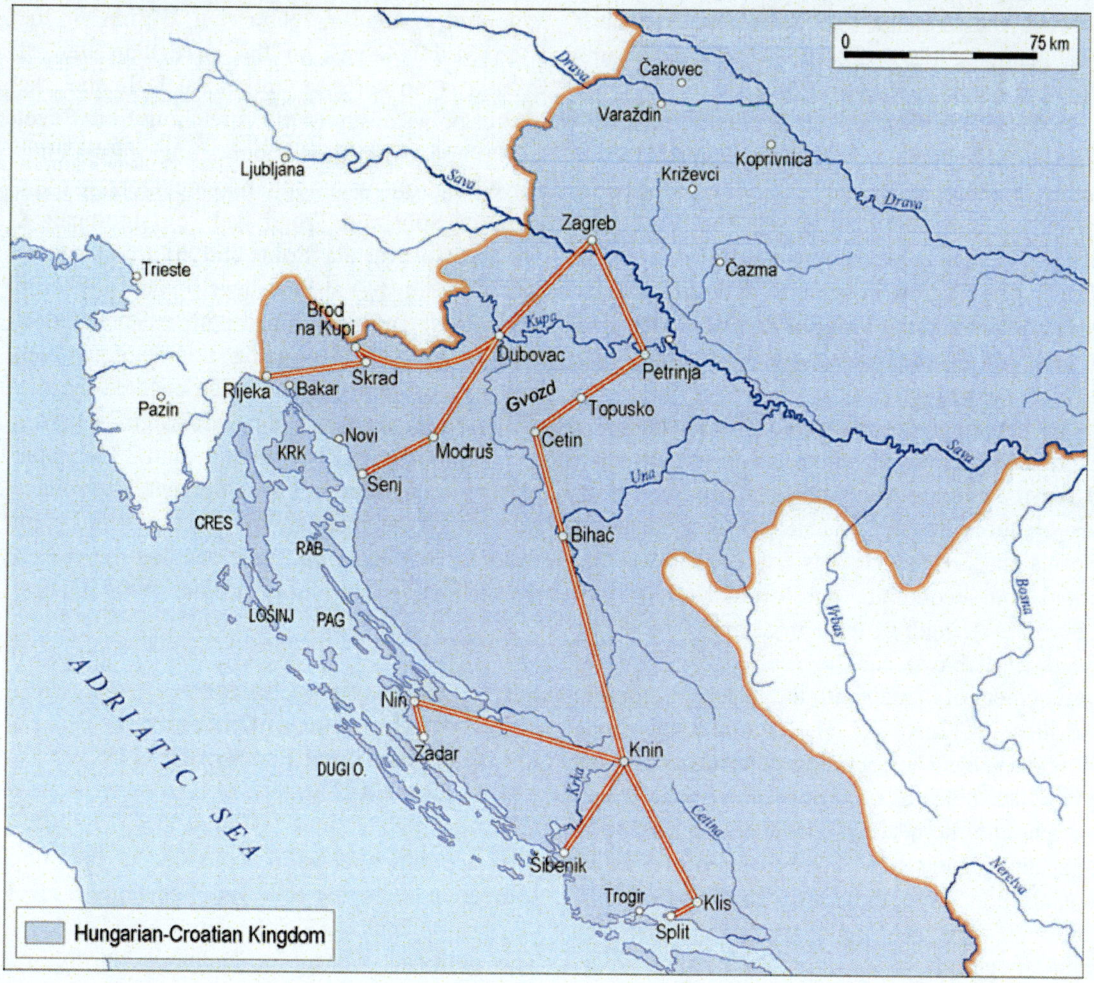

Fig. 5.5 Main transport links of Medieval Croatia

5.3 Adriatic Croatia South of Gvozd

The area south of Gvozd was outside of the strong influence of Hungarian royalty. This can be seen in the fact that there were no županijas like in Slavonia, nor royal estates or even a taxation system. Croatia south of Gvozd was characterised by the development in line with feudal particularisation, which was based on the absolute economic leadership of feudal magnates and their right to collect taxes. The bearers of the new socio-economic and administrative organisation were the nobles whose power was based on military and economic dominance. Agriculture still dominated in the total economic structure. Although herding was still the main activity, crop cultivation was increasing in importance throughout the era.

The strengthening of aristocracy and feudal social relations began at the end of the Early Middle Ages, and at the beginning of the twelfth century, some nobles became independent princes. Thereafter, they began to subjugate the rural municipalities under their leadership and turn free peasants into serfs. Due to the difficult opportunities under conditions of weakly developed agriculture on karst, free peasants would voluntarily enter into vassal service, thus supporting the process of feudalisation. The fact is, however, that the process of feudalisation did not develop in the well-developed agriculture area between the Zrmanja and Krka Rivers, where rural municipalities successfully resisted pressure from the nobility (Raukar 1997).

In the area of Kvarner, crop cultivation developed strongly in the most valuable areas of flysch zones. Herding, on the other hand, took place in the hilly and mountainous zones. The Frankopan dynasty (Croatian princes who claimed to be of Roman origin) grew in strength in the area of the island of Krk and the Vinodol Region. In the area of Senj and Modruš, castles with accompanying villages popped up.

The mountainous belt was characterised by expressly dynamic development in eastern Lika (Krbava Region)—the traditional summer grazing pastures in the transit corridor between Bosnia and Kvarner. In this era, Krbava was the population, economic, and political centre of Lika. The strength of the Kurjaković family (the rulers of Krbava) can be seen in the planned resettlement of herders into the Velebit Littoral in the area of Karlobag—a city under their rule (Rogić 1958). The founding of a diocese in the area is an indicator of stronger economic development and the need to introduce a higher level of spatial organisation. The bearers of the feudal organisation were the Gušić-Kurjaković noble line who, like the Frankopan line, emphasised their (alleged) Roman origin. Their rise to power was based on their control of a strategically important route in the eastern part of Lika (Krbava), where transport between the western part of Bosnia and Adriatic cities (Senj, Jablanac, Karlobag, and Obrovac) took place. Along with the strengthening of crop cultivation within the economic structure, the weight of the population and economic activity began to shift to the Velebit Littoral in the central part of Lika[5] (Rogić 1972a; Pejnović 1985).

In northern Dalmatia, the most agriculturally developed areas like the Ravni Kotari Region and the Krka Valley had the role of centres and were also a means of maintaining relations with neighbouring Bosnian Završje. Therefore, the rise of the Bribirski (Šubić) family, who also temporarily ruled neighbouring Bosnia in the High Middle Ages, in this particular area of Croatia is no surprise. In central Dalmatia, the area of poljes around Split, stretching from Seget to Stobreč, with their mountainous hinterlands played the role of agricultural incubators, and the cities of Split and Trogir had key significance. The areas of Kaštela, Solin, and Split poljes were characterised by strong development of crop cultivation. This, combined with the practice of herding in the surrounding mountainous belt, meant significant income which enabled the rise of the communes of Split and Trogir, ensuring their prominent political role in central Dalmatia. Apart from agricultural development, importance was placed on maintaining trade relations with Bosnia. This economic base enabled the rise of the Nelipčić noble family (Rogić 1990, 1991).

5.3.1 Dalmatia—Changes in the Spatial Concept; Urban and Economic Continuity

The second spatial whole consisted of booming Dalmatian cities that achieved their highest level of development during the High Middle Ages. The political concept of Dalmatia during the Early Middle Ages was based on five coastal (Kotor, Dubrovnik, Split, Trogir, and Zadar) and three island (Osor, Krk, and Rab) urban communities under the Byzantine rule. In the High Middle Ages, this grew to include other coastal cities—Nin, Biograd, Šibenik, Omiš, Hvar, Korčula, and Ston. Individual parts of Dalmatia were part of the Hungarian–Croatian state, while others were controlled by the Venetian Republic, which was a player in the battle for control of the eastern Adriatic coast during that era.

Thereby, coastal cities emphasised the political continuity of Dalmatia, as a region that Byzantine emperors had left to be administered by Croatian kings, with the caveat of respecting the political autonomy of the cities themselves. In contrast, the Venetian Republic also made its mark as a successor to Byzantium in the Adriatic Basin. In accordance with this, Venice used the

[5]The numerous fortifications in central Lika from this era give further evidence of this.

term Dalmatia to describe the majority of its territory on the eastern Adriatic coast, apart from Istria. The Venetian concept of Dalmatia included individual cities and islands (Cres, Lošinj, Krk, Rab, Pag, Silba, Ugljan, Brač, Hvar, Vis, Korčula, and Mljet) which they succeeded in conquering during the High Middle Ages. As Venetian influence spread into other areas of the eastern Adriatic coast, so too did the concept of Dalmatia spread geographically. Dalmatian cities were firmly integrated with their agrarian surroundings, in which advanced Mediterranean agriculture developed, thus becoming centres for the surrounding agricultural areas.

The cores of this agrarian area were the former Roman *agri,* and the development of coastal cities depended on the continuity of their maintenance. These urban centres, with the fertile workable soil in their vicinity, offered a superior basis for development. An advanced cultural landscape was shaped in the direct vicinity of coastal cities, characterised by intensive Mediterranean crop cultivation and a high level of agrarian valorisation oriented toward the cultivation of grapevines, olives, and grains (wheat, barley, spelt, and millet). The basic role of the agricultural surroundings was to supply their cities with agricultural products. Numerous agreements made between landowners (the Church, aristocracy, and citizens) and the peasants who worked their land give evidence of the dominance of grape and olive cultivation. From this, it can be concluded that these agrarian surroundings were not able to satisfy the needs of their cities in terms of basic food products.

Some local documents also bear witness to this and show how, in the sixteenth century, the Split commune[6] only produced enough grain for one-third of its total needs, while producing wine in sufficient quantity to export. Due to this, the export of grain from Dalmatian cities was forbidden, and all grains were offered for trade to ensure sufficient supply. Although the majority of communes encouraged the cultivation of grains by legal means—by limiting the amount of vineyards—the results of such policies were significantly lacking (Jelaska 1985). Coastal communes influenced the development of the cultural landscape of the nearby islands that were part of their circle of influence. Herding and forestry were developed on these islands, along with Mediterranean crop cultivation. Conserved historical documents clearly show the example of how Split (one of the most developed communes) intensively influenced the economic development of the islands of Šolta, Brač, and the eastern part of Čiovo (Novak 1957).

The agricultural surplus was trade material for Dalmatian city markets and was exchanged for animal husbandry products or exported to other Mediterranean cities, primarily those located on the Apennine Peninsula. The majority of Dalmatian cities allowed livestock and meat to be imported freely from the surrounding area,[7] indicating a shortage of meat and meat products (Jelaska 1985). Dubrovnik was in a worse situation regarding the agricultural production capabilities of its surrounding area, due to the lack of a significant amount of fertile soil. Dubrovnik was, therefore, more strongly oriented toward trade from its very founding, and was forced to conquer nearby areas in which better agricultural development was possible (the Elafiti Archipelago, Konavle Region, Pelješac Peninsula, and the islands of Mljet and Lastovo).

Dalmatian cities, primarily those that were directly on the coast, were closely linked to their natural hinterland, i.e. the Hungarian–Croatian Kingdom, in an economic sense. Export of products, among which raw materials were predominant (livestock, animal husbandry and beekeeping products, fur, and wood), from the interior took place at Dalmatian harbours, as well as the import of goods which were traded primarily to satisfy the needs of the interior (mostly salt, grain, and various tradecrafts). Regardless of strong economic integration with the wider hinterland, which belonged to the Hungarian–Croatian

[6]The Split commune was one of the most economically important in this era in Dalmatia, and boasted agrarian holdings of significant size.

[7]Customs were paid only when a good came from another commune or from within a city itself.

Kingdom, Dalmatian cities were only politically integrated into its framework for a short period.

The reason for this is the autonomous nature of these cities, which allowed for self-government, unhindered trade, and the creation of trade agreements with other autonomous Mediterranean communes. Due to this, the concepts of Croatia and Dalmatia became separate in a political sense, although the societal differences between Dalmatian and Croatian cities were minor. Documentation shows how their population, including the aristocratic and clerical classes, had already been "croaticised" by the beginning of the High Middle Ages (Mažuranić 1975; Šimunović 1997). This had been achieved largely through marriages. The chronicler Thomas the Archdeacon also wrote on this topic using the example of Split, as well as agreements regarding mutual aid among individual Roman families in cities with Croatian families that lived in the surrounding villages (Novak 1957). Additionally, Dalmatian cities were small enclaves, in terms of space and population, between the sea and the hinterland. The periods of their economic advancement demanded more and more manpower in different professions (craftsmen, merchants, sailors, soldiers, public, and household servants), which was resolved by drawing an influx of migrants into the cities and putting them to work.

Dalmatian cities were of great importance to Croatian territory because they were bearers of cultural continuity and social and economic development, and they had the role of incubators that spread Mediterranean cultural influence. Simultaneously, the bulk of the population on the nearby islands was concentrated in rural settlements and was thus largely a rural society. Island settlements were mainly situated in the interiors of the islands, where conditions for agriculture were good. An inland location also provided significant protection from piratical raiding (Raukar 2007). Significant development of island settlements began after the Venetians gained firm control of the area. The exception was the town Hvar (on the island of the same name), which had already asserted its position as the leading island settlement for the greater island area (Hvar, Brač, and Vis) in the thirteenth century.

5.3.1.1 The Venetian Republic—An Emerging Spatial, Economic, and Political Power in the Adriatic

The Republic of Venice became one of the most powerful forces in the Mediterranean during the High Middle Ages. It was founded as a typical thalassocracy,[8] and its economic, political, and military power was based on its wealth, obtained through vigorous trade with the Asian states. Its mercantile domination was inseparable from its status as the most powerful force in the Mediterranean. In order to maintain their economic strength, it was vitally important for the Venetians to keep control over their holdings on the Mediterranean coast, especially in Asia, Africa, Greece, and Crete. Therefore, the spatial expansion of the Venetian state was oriented toward those parts of the Mediterranean, where they organised their merchant emporiums. The strategic significance of the Adriatic sea is reflected in the fact that control over the Adriatic Basin enabled control of a transport corridor that linked Southeast Europe and Southwest Asia with Central Europe. In this way, the eastern and western coastlines of the Adriatic Sea, from the Venetian viewpoint, also represented an area essential for monitoring the aforementioned transport corridor. This made political and military control of both shores of the Adriatic essential (Selva 2006).

During the High Middle Ages, Dalmatian cities became objects of increasing political attention of the Republic of Venice, which the Arpadović dynasty was unable to meaningfully resist. The best evidence of the Hungarian–Croatian kings' weakness in this regard was their inability to resist the fourth crusade, which led to the loss and sacking of Zadar in 1202 (this same army would go on to sack Constantinople 1204) (Raukar 1997). It should be noted that the entire crusader army was excommunicated by the Pope for the sacking of Zadar, though the ex-

[8] Regarding the Venetian view of the Adriatic Sea, we can look at the toponym "Gulf of Venetia" (*Sinus Venetus*, *Golfo di Venetia*, *Golfo di Venezia*, and *Golfe de Venise*) which was present on Medieval Venetian maps, replacing the term *Mare Hadriaticum* (Adriatic Sea).

communications were later forgiven for all non-Venetian crusaders (Runciman 1975). The roots of Venetian aspirations for control over Dalmatian cities were not based on their humble local mercantile influence, nor their function as trade nodes between Croatia, Bosnia, and the Apennine Peninsula. Control over the eastern Adriatic was just one piece in the puzzle that would give Venice greater control over intercontinental trade,[9] i.e. between the Oriental world and Europe.

Namely, Venetian merchant ships brought large amounts of precious goods, produced in South and Central Asia, such as cloth (especially silk), spices, porcelain, and tea, from the Near East to Europe (Gresh et al. 2006). Dalmatian cities were only marginally included in this vast trade system, of which only Dubrovnik took part in a significant way. The political aspirations of the Republic of Venice for control over the coastal communes on the eastern coast of the Adriatic were linked to strategic factors, the most important of which was their location along the key geographical transit corridor between Venice and Southwest Asia. Namely, for trade between Adriatic harbours and other parts of the Mediterranean, the system of canal routes was key. This system was based on three elements:

1. At the entrance to the Adriatic from the Ionian Sea, winds and sea currents enable ships to easily sail along the eastern Adriatic coast toward the northern part of the Adriatic. As the main sea currents that enter the Adriatic from the Ionian sea move along the eastern Adriatic coast toward the north, ships were able to more easily sail toward Venice.
2. Easier maritime navigation along the eastern Adriatic coast is possible thanks to the indented coast and the great number of islands. This made orientation on the eastern coast much easier and, due to this, there were also a number of different routes, which was not the case on the western coast.
3. The ability to find shelter in the event of bad weather is significantly greater along the eastern coast, because of the many coves, bays, and inlets, whereas the western coast is much less indented. Protection from the fierce bora wind was especially important for ships, as it was one of the main causes of shipwrecks, stranded crews, and loss of valuable cargo (Rogić 1953, 1990).

The entrance and exit points were the most strategically significant parts of the aforementioned system of sailing channels. In the south, this was Dubrovnik and Korčula (on the island Korčula), and in the north, Zadar, Osor (on the island Cres), and Pula (on the Istrian Peninsula). Due to this, the largest harbours at these entry/exit points gained a political role, along with their existing strategic role. This mainly refers to Dubrovnik (which was the political centre of Upper Dalmatia during the Venetian rule from 1204 to 1358) and Zadar (which was the administrative centre of Lower Dalmatia). Until the tenth century, Venice and Genoa were the main merchant middlemen between the Near East and Europe. Starting at the same time, Venetian leaders began to systematically and determinedly push their agenda in the eastern Adriatic, not giving up despite major military defeats at the hands of Genoa and its allies, Croatian, and Hungarian–Croatian leaders.[10] The Venetian policy of limiting, or totally revoking, the autonomy of Dalmatian cities in order to quash independent economic policies, worsened political relations (Rogić 1990).

The Republic of Venice was primarily a maritime merchant power that regarded all maritime merchant cities as possible competition—that needed to be removed. Adriatic cities under Hungarian–Croatian rule enjoyed considerable economic and political freedom as the kingdom's political focus was primarily on the continent, rather than on maritime dominance. Therefore, the resistance to Venice on the part of Dalmatian cities, whereby they turned to Hungarian–Croatian kings and local Croatian aristocrats,

[9]This integral element of the Venetian trade was also called "the great world market".

[10]The Battle of Korčula (1298) and the War of Chioggia (1378–1381) were the best examples of cooperation in the battle against the Republic of Venice.

especially those who ruled over lands in the direct vicinity of coastal cities, is not surprising. In the period following the Peace of Zadar (1358), when coastal cities and the hinterland were also politically integrated, Dalmatian communes enjoyed the greatest level of development under conditions of total autonomy (Novak 1957).

The Dubrovnik, which came under Hungarian–Croatian control in 1358 and subsequently flourished economically, is the best example of this (Carter 1994). The Venetian conquest of coastal cities, primarily Dalmatian, would result in changes in their physiognomy that were visible in their architecture. The political separation of cities from their hinterland was morphostructurally expressed in the construction of stronger systems of fortification (Raukar 2007). Additionally, the establishment of the Venetian domination in the Adriatic had a positive impact on the stability of political conditions along the eastern Adriatic coast and encouraged the development of settlements on the islands via a policy of colonisation. The Venetian leadership also started settlement and economic development of more weakly settled islands[11] (Rogić 1972b).

5.3.2 Dubrovnik

Although Dubrovnik nominally acknowledged the rule of Hungarian–Croatian kings and called on them when threatened by Venice, Bosnia, or Serbia, at the beginning of the fourteenth century, it had already begun to shape the territory of what would become the Republic of Ragusa (Dubrovnik). The core of the budding Republic was formed in the area under Dubrovnik's control, extending from the area north of Zaton to the southern edge of Župa Bay and Cavtat.[12] This is the oldest holding of the Dubrovnik commune, marked by the name *Astarea* (*stereá* means "firm land" in Greek), and had already come under their control in the Early Middle Ages.

Stable Mediterranean crop cultivation developed in this area, dominated by vineyards, and the Croatian word *vinogradi* (vineyards) came to be used as a synonym for Astarea. In the thirteenth century, Dubrovnik claimed control of some of the nearby islands.[13] The islands in question were Koločep, Lopud, and Šipan, which were referred to collectively as the "three islands" and would remain a constitutive part of the Republic of Ragusa (Dubrovnik) until its final days. Until the end of the Early Middle Ages, Dubrovnik was a part of Byzantine Dalmatia and passed from Byzantine and Venetian control to Croatian, and later to Hungarian–Croatian kings. Although the beginning of the High Middle Ages was marked by the removal of Byzantine and Venetian rule and Hungarian–Croatian control over the city,[14] Dubrovnik managed not only to keep its autonomy, but develop it thoroughly.

From the start of the thirteenth century, Dubrovnik fell under Venetian rule, which limited the city's political and economic autonomy. With approval from Venetian authorities, Dubrovnik began to widen its territory in order to surveil the system of sailing channels. Although the attempt to conquer the islands of Korčula, Lastovo, and Mljet in 1240 failed, Lastovo—the furthest and most indefensible island—fell to Dubrovnik in 1252. The most important geostrategic move made by Dubrovnik was the inclusion of the Pelješac Peninsula into its territory. This happened in 1333 when the king of Serbia (the *de jure* ruler of Pelješac) and the king of Bosnia (the *de facto* ruler of Pelješac) gifted the peninsula to Dubrovnik as a sign of thanks for its peaceful diplomacy during the War of Hum.

[11]The best examples of this are salt panning on Pag and fishing on Dugi Otok (Sali).

[12]In the very beginning, when the Republic did not yet control all of Župa, Cavtat did not have a land link with the remainder of communal territory.

[13]Although they probably came under the city's control at the end of the Early Middle Ages (ninth–tenth century).

[14]Along with the short period of rule by Norman kings from southern Italy.

Taking control of Pelješac enabled Dubrovnik to surveil the Pelješac Channel, which was one of the key points of channel maritime transit along the eastern Adriatic coast. Due to this, Dubrovnik's leaders founded a new settlement called Orebić on the shore of the Pelješac Channel to entrench their control over the area. Orebić would quickly become the most important administrative and maritime centre of the peninsula, surpassing Ston. The fortification of the Ston Isthmus with massive walls and the fortification of both Ston and Mali Ston additionally secured the peninsula. The confiscation of land on Pelješac by Dubrovnik's rulers marked a period of stronger agricultural production, based on animal husbandry, crop cultivation, and vineyards. Additionally, the following maritime activities were also strengthened: sailing, fishing, salt mining, shipbuilding, and the first forms of farming of the sea (shellfish) (Glamuzina 2009).

Repeated attempts of the conquest of the island of Mljet took place in 1345 when the island was taken by the Serbian king who, after including Zahumlje in the Serbian state, became the nominal ruler of the island. In this way, during the fourteenth century, the Republic of Ragusa (Dubrovnik) had totally entrenched its position of surveillance over the system of sailing channels, which would give it an important political and strategic role in the Adriatic Basin. By cleverly exploiting weaknesses of the central governments of neighbouring states, Dubrovnik's leaders overtook strategically important parts of the nearby hinterland from Župa to Rijeka Dubrovačka in 1357.

The Treaty of Zadar (1358) freed Dalmatian cities, including Dubrovnik, from the Venetian administration and gave them to the Hungarian–Croatian king. However, Dubrovnik became further separated from the main body of Hungarian–Croatian territory by the expansion of the Kingdom of Bosnia toward the Adriatic sea. At the end of the century, Dubrovnik's cunning diplomatic efforts had taken full advantage of the feudal anarchy in the neighbouring Bosnian state that followed the death of the powerful King Tvrtko I of Bosnia. Attempts to gain control of *Primorje*[15] were successfully carried out in 1399, when they were gifted to Dubrovnik by the Bosnian king.

After Venice bought the "right" to occupy Dalmatia at the start of the fifteenthth century, Dubrovnik declined to recognise the renewed Venetian rule. In this case, Venice took no military action against the city, as it did against other Dalmatian cities that refused to recognise Venetian supremacy. The Dubrovnik commune would be the only Dalmatian city to remain under the Hungarian–Croatian crown and would continue to widen its territory even in times of crisis. Dubrovnik briefly gained control over the islands of Korčula, Hvar, and Brač, from 1413 to 1417, due to an order from the Hungarian–Croatian king who was feuding with Voivod Hrvoje Vukčić Hrvatinić—the previous ruler of the three aforementioned islands. Also, after a series of nonviolent political and diplomatic efforts starting the mid-fourteenth century, the area from the Konavle Region (to the southeast of Dubrovnik) was taken.

The area was transferred to Dubrovnik in 1419 according to the treaty, whereby the Bosnian noble Sandalj Hranić surrendered the southeastern part, and the brothers Petar and Radoslav Pavlović the northwestern half in exchange for money and other compensation.[16] This agreement was affirmed by the reigning king of Bosnia. It was not until 1427, however, that Dubrovnik actually took control of the part surrendered to them by the Pavlović brothers. This would complete the spatial formation of Dubrovnik's territory. Dubrovnik led all parts of its territory with a firm hand, via dukes on the islands Lopud (Koločep was also under the same jurisdiction), Šipan, Lastovo, Mljet, Pelješac Peninsula (the seat of which was Ston),[17] Astarea (the seat of which was Donja Čibača in Župa),

[15]*Primorje* (the Littoral) is known today as *Dubrovačko primorje* (Dubrovnik Littoral), but then it was called either *Slano primorje* (Slano Littoral) or *Terra Nova*.

[16]This also included acceptance into the ranks of Dubrovnik's nobility for the entire male line.

[17]Trstenica, the northwestern part of the peninsula, and Janjina were part of a special administrative unit under the rule of a captain.

the Dubrovnik Littoral (Slano), and in the Konavle Region (Pridvorje).[18] Although Dubrovnik was first called "Republic" in one internal government document in 1441,[19] the city still officially recognised the authority of the Hungarian–Croatian crown. Additionally, Dubrovnik consistently confirmed that it belonged to Dalmatia as an integral part of the Hungarian–Croatian state. However, the city commune functioned completely independently, especially after its liberation from the Venetian authorities (Foretić 1980).

Intermittent conflicts, with leaders in the direct hinterland of the Republic of Ragusa (Dubrovnik) during the second half of the fourteenth century, resulted in the near-total destruction of all elements of the cultural landscape including settlements and agricultural land (Carter 1994). These problems were gradually resolved during the following centuries, after the Ottomans took Bosnia and Herzegovina, and it seemed as though Dubrovnik would soon find itself under their protection. After Dubrovnik came under Ottoman protection, the city's leadership stopped referring to Dubrovnik as a constituent part of Dalmatia in its internal documentation. The main reason for this was that being part of Dalmatia automatically meant an acknowledgement of the sovereignty of Dubrovnik's main rival in the Adriatic: Venice, which controlled the bulk of Dalmatia (Fig. 5.6).

5.3.3 Istria—A Political and Cultural Contact Area

Until 1230, Istria remained a part of the German state, but from 1234 to 1374, it fell under the administration of the Patriarchy of Aquileia. During German rule, a network of administrative centres which were administered by local administrators (*kastelan* or *gastaldi*) was formed, while Gafers had the role of the main centre (contemporary Koper in Slovenia[20]). During the High Middle Ages, the interior of Istria came under the leadership of the Habsburg dynasty after 1374, as did nearby Rijeka (from 1471 onward). With the Habsburg rule, Germanic influence continued, which enabled the application of advanced agricultural techniques and the administration of feudal holdings.

Thereby, the settlement of the Germanic population into cities reflected positively on the development of crafts and trades. This area was included within the borders of the Duchy of Pazin, where the application of agricultural innovations from Central Europe resulted in stronger development of crop cultivation. In contrast, the coastal part of the Peninsula was under the rule of Venice during the fourteenth century, and the area was strategically significant for its role in overseeing maritime trade in the eastern Adriatic. This meant that Istria was divided between two powerful political entities. This divide would persist until the end of the eighteenth century and the fall of the Republic of Venice. Venetian conquest was followed by fierce armed resistance on the part of Istrian communes, which were aided by units from the Patriarchate of Aquileia, the Austrian duke, and the Croatian Frankopan family. War damage during the fourteenth century, which was followed by a plague epidemic, led to population regression and deterioration of the cultural landscape in rural areas[21] (Bertoša 2007).

The lack of population allowed the Venetians to enact settlement of abandoned areas after conquering the western coastal areas of the peninsula in the second half of the fourteenth century. With this goal in mind, Venetian authorities gave settlers abandoned land to use in perpetuity with the condition that they plant crops within the first five years after taking possession. Simultaneously, settlers were freed

[18]Because Cavtat was under the administration of a Captain.

[19]In the same document the concept of "Republic" is used simultaneously as a synonym for the Dubrovnik commune.

[20]The significance of Kopar is clearly shown by the toponym Capodistria, which is taken from the Venetian name of Caput Histriae (capital of Istria).

[21]In the area of southern Istria, 80% of villages became abandoned by the end of the sixteenth century.

Fig. 5.6 Spatial development and expansion of the Republic of Ragusa (Dubrovnik) (*Sources* Regan 2003; Dugački and Regan 2018)

from all taxation and civil service responsibilities for a period of 20 years. Venetian leadership also freed all settlers in Istrian cities and villages from taxation and fees for a period of five years. Regardless of the undertaken measures, settlers were not able to create significant crop yields and ended up re-settling in Istrian cities. By 1421, the Venetians had totally broken all Istrian communes of the will to resist and ruled the majority of the Peninsula (Merlić 2008).

The exception was the interior, which continued to be characterised by Central European feudalism, with alternating periods of advancement and regression. Namely, conflicts between the Habsburg and Venetian administrations, as well as conflicts between individual communes, reflected negatively on settlements in the interior of Istria. Intermittent plague epidemics also took place, and the flood plains and swampy areas along the Mirna and Raša Rivers were malarial hotbeds (Rogić 1975). The process of slavicisation of rural areas, primarily in the interior of Istria, was mainly finished during the High Middle Ages and was much more complete than in cities. This came about partially due to the strengthening of feudal social relations. Namely, German aristocrats colonised the Slavic population, from which the majority of serfs working in the fields were drawn, according to their plans.

5.4 Historical–Geographical Developments of Neighbouring Lands in the Middle Ages

5.4.1 The Geographical Nucleus and Spatial Developments of the Bosnian State

During the High Middle Ages, individual parts of contemporary Croatian territory were, for shorter or longer stretches, included in the territories of neighbouring countries. Economic, political, and military strengthening was followed by spatial expansion of neighbouring countries into Croatian territory, while times of crisis were followed by opposing processes. Among neighbouring states, Bosnia is important to mention, unlike Medieval Serbia, which only briefly interacted with Croatia in this era.

During the Middle Ages, Bosnia was a constant neighbouring presence for Croatia, and its link to Croatia can be seen in economic and political ties. The Bosnian state formed its core in the central Dinarides in the Early Middle Ages, at the upper flows of the Bosna and Vrbas Rivers and the middle flow of the Drina River, which became the geographical framework for the political and territorial organisation.[22] This area did not have areas of winter pastures and was, in this regard, oriented toward neighbouring countries from its inception as a political whole. Bosnian herders were primarily oriented toward the area of Croatia for winter pastures.

The political ties between Bosnia and the Hungarian–Croatian Kingdom were based on economic relations—Bosnia would become a part of the Hungarian–Croatian Kingdom in the form of an autonomous Banovina starting in the eleventh century. When Tvrtko I came to power in 1377, Bosnia became a kingdom. The link between the Bosnian and (Hungarian–) Croatian kingdoms can be seen in ethnic, societal, economic, cultural, and political characteristics (Raukar 2000). The location of the first significant fortified cities and castles in the valleys of the larger rivers indicates the early valorisation of transport routes, which were already important in prehistory and Antiquity. This indicates the importance of the transverse transport routes through the Dinarides, passing through the Bosna and Vrbas River valleys and linking the Mediterranean with Central Europe and Pannonia.

The rise of the first significant urban settlements in the Bosna River Valley should be regarded in this light—Vrhbosna (located where Sarajevo is today), and the temporary royal seats (Visoko, Vranduk, Sutjeska, Žepče, Maglaj, and Doboj). The development of a mercantile emporium in the Neretva Delta by Dubrovnik, which is noted by numerous historical documents from Dubrovnik, is evidence of the transit and geographical valorisation of important transit routes in the pre-Ottoman era. Merchants from Dubrovnik were present in cities in the Bosna River Valley during this period. The inclusion of merchants from Dubrovnik in the exploitation of mining wealth, via concessions given by Bosnian kings, additionally highlights the importance of this part of the Bosnian state in economic and political terms.

The reason that Bosnia recognised the supremacy of Croatian and Hungarian–Croatian kings is economic, showing the vassal-like relationship. Additionally, individual Croatian nobles sometimes led Bosnia. The best example of this is Ban Borić (1154), the first Bosnian ban who also belonged to Croatian nobility[23] (Truhelka 1941). The dukes of the Šubić-Bribirski family had the most expressed role in this sense (Rogić 1990) and were also leaders of Bosnia in their time (in 1302 Mladen I became the Ban of Bosnia).

The beginning of the High Middle Ages marked the strengthening of Bosnian leaders, which was a sign of territorial expansion in geographical terms (Fig. 5.7). This expansion was oriented toward four neighbouring areas along the Croatian border:

1. The peri-Pannonian area of contemporary northern and northeastern Bosnia came to be a part of the Medieval Bosnian state in the thirteenth century, but was soon lost for a short period before being reincorporated again in the fourteenth century.
2. The "Lower Lands" were the central parts of the riverlands of the Sana and Vrbas Rivers as well as the area of Završje and came under the control of Bosnian leadership in the fourteenth century.
3. Northern and central Dalmatia, the Mediterranean areas with the best winter pastures (in the vicinity), were taken by Bosnian leaders in the fourteenth century.
4. The conquest of Hum (Zahumlje) and Travunia, also in the fourteenth century, was a significant event for the sub-Mediterranean belt.

[22]The first mention of Bosnia was in the tenth century under the name "zemljica Bosna" (the land of Bosnia).

[23]The family holdings of Ban Borić were in Slavonia in the surroundings of Slavonski Brod and in the Bosnian Posavina Region.

Fig. 5.7 Spatial development of the Kingdom of Bosnia during the Middle Ages (*Source* Proleksis Encyclopaedia, The Miroslav Krleža Institute of Lexicography, http://proleksis.lzmk.hr. Accessed Nov. 26, 2019)

In contrast to the neighbouring area of the Hungarian–Croatian Kingdom, Bosnia had a very weakly developed crop cultivation component in its economic structure in the High Middle Ages. The reason for this was found in the environment (primarily in climatic and pedologic characteristics), which did not offer good conditions for the development of crop cultivation, as well as the weak influence of western clerical orders (Jesuits, Templars, and Hospitallers) that spread new, advanced techniques of working in the land in their areas of influence. Due to this, transhumant herding held the main role in the Bosnian economy, and the newly conquered sub-Mediterranean area (and its winter pastures) was left to the administration of nobles who often had great power and influence.

Because of the importance of the sub-Mediterranean area, a special name for it came into use—Hercegovina—as it fell under the administration of a *herceg* (*Herzog* or duke). The need for additional pasture areas and the ability to develop the crop cultivation element in the economy dictated the direction of expansion toward Slavonia. Territorial expansion toward Croatian areas in the west was also based on the need to monitor trade (import and export), which Bosnia conducted via Dalmatian harbour cities. This shows the significance of the Adriatic coastal façade for Medieval Bosnia. The economic orientation of Bosnian territory toward the west was a consequence of its dependence on Adriatic cities, primarily Split and Dubrovnik. This was the end destination for goods coming from the Dinaric interior under Bosnian rule, and via these very trade routes, many social, cultural, and intellectual influences from the Mediterranean area, as well as Western Europe, came to Bosnia (Raukar 2000).

During the period of territorial expansion, new urban centres were founded in the valleys which served as the main trade and travel routes: Foča, Goražde, and Višegrad in the Drina River Valley; Jajce in the Vrbas River Valley; Ključ in the Sana River Valley; and Konjic in the Neretva River Valley. After the death of Tvrtko I in 1391, the Bosnian state weakened as it fell slowly into feudal anarchy, and leadership over individual Croatian territories was taken over by local aristocratic families (Rogić 1990).

5.4.2 Medieval Serbia

In the thirteenth Century, the unification of *Raška* (Rashka) and *Zeta* formed the Serbian state. The expansion of this new state during the High Middle Ages was primarily oriented toward the southeast and the Aegean and Ionian coasts. This orientation was due, in large part, to the weakening of Byzantium. The territorial expansion of Serbia toward other areas was hindered by strong neighbours to the west (Bosnia), north (the Hungarian–Croatian Kingdom), and east (Bulgaria). The Pelješac Peninsula and the Dubrovnik Littoral (Astarea) had been under the rule of Zahumlje (Zachlumia or Hum) and then Bosnia until the start of the fourteenth century when they came under the control of the King of Serbia (although the de facto ruler was the ban of Bosnia). Serbian rule over these areas was short and did not leave significant traces in the cultural landscape, as both Bosnia and Serbia had given up any claim to Pelješac Peninsula by 1333, and Astarea was ceded to Dubrovnik in 1399 (Rogić 1990).

References

Bertoša M (2007) Istarsko rano novovjekovlje: razvojne smjernice od 16. do 18. st. In: Budak N (ed) Dalmacija, Dubrovnik i Istra u ranome novom vijeku. Leykam International, Zagreb, pp 80–120

Budak N (2007) Hrvatska i Slavonija u ranome novom vijeku. Leykam International, Zagreb

Carter FW (1994) The political geography of the Dubrovnik Republic. Acta Geographica Croatica 29:77–98

Dugački V, Regan K (eds) (2018) Hrvatski povijesni atlas. Leksikografski zavod Miroslav Krleža, Zagreb

Foretić D (1980) Povijest Dubrovnika do 1808.: prvi dio – od osnutka do 1526. Nakladni zavod Matice hrvatske, Zagreb

Fuerst-Bjeliš B (1996) Historijsko-geografski pogled na pitanje prvobitne teritorijalne organizacije Hrvatske. Acta Geographica Croatica 31:115–124

Glamuzina N (2009) Pelješac. Naklada Bošković, Split

Glamuzina N, Fuerst-Bjeliš B (2015) Historijska geografija Hrvatske. Sveučilište u Splitu, Split

Gresh A, Radvanyi J, Rekacewicz P, Samary C, Vidal D (2006) „L'Atlas" du Monde diplomatique. Le Monde diplomatique, Paris

Grgin B (2012) Vrhunac srednjovjekovnog razvoja. In: Goldstein I, Goldstein S (eds) Povijest grada Zagreba – knjiga 1. Od prethistorije do 1918. Novi liber, Zagreb, pp 64–107

Jelaska J (1985) Splitsko polje za turskih vremena. Logos, Split

Kurelac M (2000) Hrvatska i središnja Europa u doba renesanse i reformacije. In: Hercigonja E (ed) Hrvatska i Europa – kultura, znanost i umjetnost: svezak II. Srednji vijek i renesansa (XIII–XVI. stoljeće). Školska knjiga, Zagreb, pp 39–60

Laci S (1982) Razvoj naseljenosti Međimurja. Geografski glasnik 44:51–68

Magocsi PR (2002) Historical atlas of Central Europe: from the early fifth century to the present. Thames & Hudson, London

Mažuranić V (1975) Prinosi za hrvatski pravno-povjestni rječnik. Informator, Zagreb

Merlić E (2008) Buzeština kroz povijest. Vlastita naklada, Pazin – Buzet

Mohorovičić A (2000) Razvoj naselja i gradova na području sjeverne Hrvatske. In: Hercigonja E (ed) Hrvatska i Europa – kultura, znanost i umjetnost: svezak II. Srednji vijek i renesansa (XIII–XVI. stoljeće). Školska knjiga, Zagreb, pp 553–583

Novak G (1957) Povijest Splita: knjiga prva – od prethistorijskih vremena do definitivnog gubitka pune autonomije 1420. godine. Matica hrvatska, Split

Pejnović D (1985) Srednja Lika: socijalnogeografska transformacija. Centar za kulturu – Muzej Like, Gospić

Raukar T (1997) Prostor i društvo. In: Supičić I (ed) Hrvatska i Europa – kultura, znanost i umjetnost: svezak I. Srednji vijek (VII–XI. stoljeće) rano doba hrvatske kulture. AGM, Zagreb, pp 181–195

Raukar T (2000) Hrvatska na europskom prostoru. In: Hercigonja E (ed) Hrvatska i Europa – kultura, znanost i umjetnost: svezak II. Srednji vijek i renesansa (XIII–XVI. stoljeće). Školska knjiga, Zagreb, pp 5–36

Raukar T (2007) Studije o Dalmaciji u srednjem vijeku: odabrane studije. Književni krug, Split

Regan K (ed) (2003) Hrvatski povijesni atlas. Leksikografski zavod Miroslav Krleža, Zagreb

Rogić V (1953) Senj: prilog poznavanju položaja i regionalne funkcije. Geografski glasnik 14(15):47–64

Rogić V (1958) Velebitska primorska padina: prilog poznavanju evolucije krškog pejzaža (nastavak). Geografski glasnik 20:53–110

Rogić V (1972a) Primjer geografske analize strukture naseljenosti gorsko-planinskih regija: ličko goranski prostor. In: Panov M (ed) Zbornik na Jugoslovenskiot simpozium za problemite na selskite naselbi i zemjodelskoto proizvodstvo, Ohrid, May 1972. Sojuz na geografskite instiucii na SFRJ – Geografsko društvo na SR Makedonija, Skopje, p 96

Rogić V (1972b) Regionalno-geografski aspekt prostora paške komune. Geografski glasnik 33–34:141–157

Rogić V (1975) Osnove historijsko-geografskog razvoja. In: Cvitanović A (ed) Geografija SR Hrvatske – knjiga V: Sjeverno hrvatsko primorje. Školska knjiga, Zagreb, pp 42–45

Rogić V (1990) Regionalna geografija Jugoslavije, knjiga 1 – prirodna osnova i historijska geografija. Školska knjiga, Zagreb

Rogić V (1991) Nacrt uvjetno-homogene regionalizacije graničnog dijela srednjodalmatinskog prostora: primjer teritorija trogirske općine. Radovi 26:35–46

Runciman S (1975) A history of the Crusades III: the Kingdom of Arce and the Later Crusades. Cambridge University Press, Cambridge

Selva O (2006) L'immagine della Dalmazia nella cartografia antica. In: Rossit C, Selva O, Umek D (eds) Imago Adriae – L'Adriatico s l'Abruzzo nelle antiche carte geografiche. SIGRAF, Pescara

Šimunović P (1997) Onomastička svjedočanstva nakon doseobe. In: Supičić I (ed) Hrvatska i Europa – kultura, znanost i umjetnost: svezak I. Srednji vijek (VII–XI. stoljeće) rano doba hrvatske kulture. AGM, Zagreb, pp 401–414

Truhelka Ć (1941) Studije o podrijetlu: etnološka razmatranja iz Bosne i Hercegovine. Matica hrvatska, Zagreb

Vujić A (ed) Proleksis Encyclopaedia. The Miroslav Krleža Institute of Lexicography. http://proleksis.lzmk.hr. Accessed 26 Nov 2019

6 Ottoman Conquest and Its Impact on Territorial Reduction and Spatial Change of Croatian Territory

Abstract

In Croatian territory, the Late Middle Ages were characterised by spatial and societal disintegration processes. This mainly coincided with the era of the Ottoman expansion—a new power that straddled three continents at the apex of its power. The general crisis in Europe worked to the advantage of the Ottoman Empire, which was more advanced than Christian European states in almost every respect. The Ottoman expansion would bring about a radical change in the cultural landscape of Croatian territory. On the one hand, the border areas with the Ottoman Empire would be marked by the complete destruction of all elements of the cultural landscape as a result of continual raiding and battle. The most prominent destructive processes would be the agrarian population discharge from the border areas and the disappearance of settlements. The territorial advancement of the Ottoman Empire, generally from the southeast to the northwest, induced great migration waves toward the more secure lands in the northern and western parts of contemporary Croatia and Austria, or even overseas. The borderlands were emptied of most of their population, and came to be occupied only by semi-nomadic Vlach warriors–herders. Areas incorporated into the Ottoman Empire, mostly in the eastern parts of the Pannonian Basin, were included into the Ottoman administrative system of eyalets and sanjaks and developed Oriental-Islamic types of settlements, although their cultural influence would be largely eradicated after the Ottoman retreat in later centuries.

Keywords

Late Middle Ages · Croatia · Ottoman Empire · Republic of Venice · Vlachs · Sanjak · Eyalet

B. Fuerst-Bjeliš and N. Glamuzina, *The Historical Geography of Croatia*,
Historical Geography and Geosciences, https://doi.org/10.1007/978-3-030-68433-4_6

The Castle of Ogulin (1500): a Renaissance fortification in Gorski Kotar
The Castle of Ogulin was built by the powerful Croatian aristocratic family Frankopan as one of their military strongholds during the era of the Ottoman conquests. It became the seat of the important 13th Regiment of the Croatian Military Frontier. The Castle was the nucleus around which the settlement of Ogulin would grow (Photograph: Barbara Bjeliš. Reproduced with permission)

In Croatian territory, the Late Middle Ages were characterised by spatial and societal disintegration processes—this was also the case in other parts of Southern and Southeast Europe. In the area of Southeast Europe, the Late Middle Ages mainly coincided with the era of Ottoman expansion—a new political, military, and economic power that straddled three continents at the apex of its power.

6.1 The Medieval European Feudal Crisis as a Prerequisite of Spatial Disintegration

Political and economic conditions in the Hungarian–Croatian Kingdom were not so different from those in other European states of the day. The end of the High Middle Ages marked the high point of the crisis of the political, social, and economic systems. The driving force behind said crisis was feudal particularism, which manifested spatially as political fragmentation and socially as increasing feudal anarchy. Feudal particularism resulted in frequent battles for power, which sometimes became full-blown civil wars. Additionally, many feudal lords resisted any strengthening of central governmental structures because they feared that their political and economic power might be limited or broken. Furthermore, feudal particularism reflected poorly on the economic system because feudal lords monopolised all income generated on their lands. Trying to squeeze every last bit of income from their holdings, they continually increased levies, which usually led to peasant revolts and conflicts with urban citizens. The pillars of resistance to feudal oligarchs would be the very cities that were burdened by numerous levies, taxes, and customs. The societal crisis also deepened, and faith-motivated persecution of Jews and other "heretical" congregations (such as patarini[1] in Bosnia) took place.

The aforementioned crisis engulfed the new economic and political core of Croatian territory in the western part of the Pannonian Basin (then referred to as Slavonia). Reasonably dynamic economic, demographic, and urban development slowed in the second half of the fifteenth century, coinciding with limitations on urban citizens' (who were the bearers of economic advancement) freedoms. This was primarily a consequence of political crises caused by changes in leadership in the Hungarian–Croatian Kingdom (the most important of which being the end of the rule of the Capetian House of Anjou) and the era of conflicting interests of the Habsburg and Luxembourg dynasties. The appearance of new capitalistic urban families (e.g. Fugger) and a new generation of feudal magnates followed this period of economic and social stagnation. The lower levels of the aristocracy lost much of their status due to the strengthening of mid- and high-level aristocrats under the protection of the Jagiellonian dynasty, which led to the weakening or disappearance of nearly all noble counties—with a few notable exceptions (Budak 2007). Feudal anarchy in Croatian territory reached its zenith between 1490 and 1520 as a result of political and economic strengthening of powerful feudal lords and their increasingly overt conflicts with lower level aristocrats. Moreover, from 1520 onward, the position of peasant subjects would continually worsen (Kurelac 2000).

Under these conditions, all attempts to strengthen central leadership and territorial organisation of the political system in Croatia failed, as was the case in neighbouring Bosnia and Serbia. There were, however, contradictory developments in the consolidation of the central government, which occurred within the two states that extended across Croatian territory. The first of these was the Republic of Venice, which occupied most of the Dalmatian coast and islands in the second half of the fifteenth century as well as the Istrian and Kvarner islands. Its strong central government depended on the network of coastal cities in which (Venetian) captains presided as representatives of local government. The only such city to have central administrative function and was not on the coast was Buzet in

[1] Also Cathars, Albigensians, Bosnian Christians, and Bogomils. These were some of the terms for dualistic denominations which taught that there were two gods: one good and the other evil. This teaching was condemned as heretical at the Third Council of the Lateran (1179).

Istria. The second state was the Republic of Ragusa (Dubrovnik), which developed as an individual political unit after the expansion of the Ottoman Empire into the territory of the Hungarian–Croatian Kingdom. Two events contributed to the political separation of the Dubrovnik commune from the rest of Croatian territory.

The first was the Ottoman conquest in the southern parts of Croatia between the Cetina River and the town Ston (1499), which physically divided the area of the Republic of Ragusa (Dubrovnik) from the rest of the Hungarian–Croatian Kingdom. The second was the political climate: after the military defeat and death of the Hungarian–Croatian king at the Battle of Mohács (1526), the leaders of the Republic of Ragusa (Dubrovnik) used the ensuing war of succession to the Hungarian–Croatian throne between Ferdinand Habsburg and Ivan Zapolja to gain further autonomy. Gradually, the Republic of Ragusa (Dubrovnik) ceased to recognise the primacy of the Hungarian–Croatian throne and formally began to function as an independent political unit under the protection of the Ottoman Empire. The Republic of Ragusa (Dubrovnik) continued to develop thanks to trade with its hinterland, as well as with other Mediterranean ports. Its territory was periodically threatened by raider attacks, i.e. by Hajduks[2] and Uskoks,[3] who regarded Dubrovnik as an Ottoman ally (Rogić 1953).

Additional problems arose from increasingly open hostility from the Republic of Venice over control of the system of sailing channels. This would be solved, however, via diplomacy. A systematic policy of centralisation was linked to the gradual reduction in the privileges of individual municipalities within the Republic. The reason behind the increasing dissatisfaction of members of lower social classes was the limitation and revocation of peasants' rights, such as the right of the serf to leave his or her piece of land and move into the city, or leave via ship, without penalty (Foretić 1980).

6.2 Ottoman Conquests in Europe

A new power was on the rise in Southeast Europe and the lands which bordered the Hungarian–Croatian Kingdom during the era of crisis in the High Middle Ages: the Ottoman Empire (Fig. 6.1). The rise of the Ottoman state can be viewed in terms of the weakness and inability on the part of European countries to mount an organised resistance, but the fact remains that the Ottoman Empire was the premier power in Europe and the Near East in this era—far stronger than its European neighbours. This superiority can be observed on several levels.

In a political sense, the Ottoman Empire had a strong and effective central government with a clear and firm hierarchical structure, as well as a system designed to protect political servants from abuse. A firm legal framework covered all citizens and servants, and no one (not even the ruler) was above the law. The head of government was the *sultan* who was aided in leadership by the *Divan*, the imperial council which was lead by the *grand vizier*. The sultan was the absolute ruler and was responsible for naming the most able and loyal servants to the highest positions, regardless of their origin.

Due to this, the influence of aristocratic circles on official policy and the sultan himself was less than in Christian European countries, and the functions of state government were more

[2] A Turkish term meaning brigand. They were armed rebels who acted in groups to attack and pillage travellers and villages. They are a relic of Medieval brigandry, and during the Ottoman area, they were a paramilitary organisation in the service of a state, i.e. Venice, and directly involved in war operations. In the folk traditions of Southeast Europe, they are celebrated as anti-Ottoman warriors and national heroes.

[3] A Croatian term meaning ambusher. They were largely autochthonous inhabitants or refugees from areas conquered by the Ottomans who were organised into paramilitary groups. During the sixteenth and seventeenth centuries, they entered conquered areas and engaged in guerilla warfare against the Ottoman rule. The most well-known Uskoks were from Senj. They combined military actions on both land and from the sea, at first exclusively against the Ottomans, but later also against the Venetians. Their style of warfare was characterised by frequent raiding and pillaging.

Fig. 6.1 The beginning of the expansion of the Ottoman Empire into Europe in the fourteenth and fifteenth centuries (*Source* Regan 2003)

efficient. The decisions of the leadership depended on a large bureaucratic apparatus to carry them out. The state bureaucracy kept a written registry and communicated primarily by writing, which also helped reduce abuses of power. The society was divided into two classes, whereby ethnicity and religion played no part: military (leadership) and *raya*[4] (civilians who paid taxes).

Raya were not bound to the land that they worked or a (feudal) lord, like European serfs, and were able to freely move from one place to another, provided that they paid the appropriate fee. Any abuse or ill-treatment of servants was harshly punished, and everyone, regardless of faith or ethnicity, could complain directly to the state council—the highest civil institution, i.e. the Divan—if their rights had been violated. The empire was divided into eyalets (*beylerbeylik* or *pashalik*), which were each overseen by a pasha (*beylerbey*), named from the ranks of military officers, who reported directly to the sultan and the Divan. Eyalets were further divided into 34 sanjaks (*sancak*—also sometimes called by the Arabic term *liwa*) under the oversight of military administrators—called beys (*sancak-bey*). As opposed to eyalets or sanjaks that were organised on a military basis, there were smaller territorial units—judicial districts (*kaza*, and later *kadıluk*)—headed by a judge (*kadı*), and smaller districts (*nahiyah*) led by judicial aids according to judicial jurisdiction. The sultan tended to move pashas and beys from one territorial unit to

[4]A Turkish term derived from Arabic (rāʿāyā) meaning subjects or commoners. In Ottoman legal classification in the classical era (fifteenth century to seventeenth century), this was the term for all Ottoman subjects who paid taxes, regardless of religion, i.e. those who were not part of the privileged military class.

another regularly, after a certain amount of time had passed. In this way, the danger that any given local leader might become too popular and consolidated in the territory that they oversaw was limited, as were abuses of power and feudal disintegration (Šabanović 1959; Moačanin 1998; Rogić 1990; Matuz 1992).

The economic superiority of the Ottoman Empire was reflected in the obligation of the individual to pay taxes, which was the sole privilege of the state government to collect. This made it impossible for individual aristocrats to increase their personal income by levying their own taxes, resulting in peasants who were much more motivated to work than their European counterparts. The main source of income was agriculture, which was organised into timars (*tımar*—based on the Byzantine *pronoia* system). These were feudal holdings owned by the government and given over to the administration of lords called timariots (*tımarlı sipahi*). These lords did not have ownership privileges on the land they administered, they simply collected rent. In exchange for becoming a timar administrator, as well as for getting individual holdings (*hâss*), they were required to serve a certain amount of time in the military. Timariots had the role of civil servants who upheld the law of the land and enforced the rulings of judges. Timars were state-owned feudal holdings, which the state had the right to take and re-assign to another timariot (Rogić 1990; Matuz 1992).

Furthermore, the leadership of timars was not hereditary. The peasants who worked on timars had the status of hereditary clients, who had the right to work some of the lands for personal use in exchange for daily labour on state-held plots. Peasants were not allowed, however, to change the way the land was used, and if a given plot went unworked for three years, it could be given to someone else by the timariot. As the state continually expanded the total amount of worked agricultural land, timariots would be rewarded for the amount of peasants who moved to the timar and expanded its amount of worked agricultural land (Inalcik 2002). In contrast to European Christians, these peasants were not serfs, and were thus not obligated to perform forced labour. Additionally, timariots did not have judicial power over them (Matuz 1992).

The religious superiority of the Ottoman Empire was reflected in its tolerance and acceptance of other monotheistic religions, which was not the case in Christian Europe at the time. All peasants living within the empire were able to practice their faith unhindered, and the largest congregations were divided according to millets,[5] of which Orthodox Christians had the most privileges. Ottoman state authorities did not engage in forced conversion to Islam, but Muslims had more privileges than other religious groups. This meant that advancement in civil service required conversion to Islam. Moreover, in contrast to Medieval Europe, Jews were saved from persecution and were often civil servants or local officials; during the fifteenth and sixteenth centuries, Ottoman leadership encouraged Jews to immigrate because they were considered skilled in matters of trade and generating wealth. The policy of religious tolerance stems from the very beginnings of the Ottoman state on the edges of Asia Minor in areas where Christian and Islamic lands met. It was here that, even in the embryonic stages of a new state, the Ottomans applied one of the most important principles of Islam: to guarantee life and property to members of other monotheistic religions along with the freedom to practice their religion and live according to their religious law—under the condition that they pay a regular tithe to the state. This was the main reason for the speedy capitulation of the population of Christian states of Southeast Europe to the Ottoman rule (Rogić 1990; Inalcik 2002).

The military might of the Ottoman Empire was based on a number of different mobile and fast soldiers and a core of Janissaries,[6] elite shock troops, i.e. heavy infantry, under the direct command of the sultan. Thanks to its disciplined Janissary army core, the Ottoman Empire was the

[5]A Turkish term, derived from Arabic (*millah*), which means "nation".

[6](*Yeniçeri*) Elite Ottoman soldiers. The term is derived from a Greek word meaning "new soldier".

first state with a standing army on European soil. The Ottomans showed a great aptitude for using firearms and new battle tactics they encountered in Christian Europe. Along with Janissaries, all other classes of society were able to enter military service, regardless of ethnicity and religion, and were thereby freed from paying taxes. The Ottoman expansion was based on relentless raiding (*akın*)[7] carried out in the territory of foreign adversaries. This was a form of warfare that the Ottomans had perfected, and the key to it lay in the deployment of irregular forces—Akinjis[8] and Martolos[9]—on the edges of the Ottoman border territory (*serhat*[10]) where they engaged in constant raiding and pillaging of neighbouring enemy territory. In fact, the main source of income for these irregulars came from raiding and pillaging, as they were not paid soldiers. Akinjis and Martolos also had the duty of collecting and forwarding information on the movements and whereabouts of enemy armies. The high mobility of Akinjis and Martolos was the basis for their ability to mount constant "lightning raids" into enemy territory. Additionally, their status as semi-legal military forces (paramilitary) allowed them to attack without consequences (to the Ottoman Empire) during times of negotiated armistice.

The Ottoman Empire exhausted the population living on its borders (especially those who actively resisted) with its unending campaign of raiding, pillaging, killing, destruction of property, and enslaving. After a certain amount of time, this invariably led to the exodus of the population in these regions, who left for more secure parts of Europe after seeing that their feudal lords could no longer protect them. Following such a population exodus, the regular Ottoman army would move in, take military fortifications, and face any remaining enemy armies on the battlefield (if needed). Once these areas had been thoroughly wrung dry of human potential and other resources, and occupied by Ottoman military units, they would be formally annexed into the empire. Thereafter, Akinjis and Martolos would move to the new border area and the cycle would begin anew. The former border area would then be populated with farmers and herders, further integrating the area into the state. Ottoman military strength was linked to the high level of motivation among its soldiers. Material motives (the spoils of war) were, alongside non-material motives (e.g. fighting for the sultan and spreading the faith), extremely important. Moreover, the ability of all distinguished warriors to obtain a timar was made into law in the second half of the fourteenth century. Even lower level soldiers were regularly rewarded with large estates (*ziamet*) (Matuz 1992). It should be mentioned that, after being conquered, the Ottomans allowed the armies and nobles of defeated lands to become part of the empire's military, regardless of whether they were Christian or Muslim.

The Ottoman Empire was also *superior in a social sense* in relation to European countries, because it was a society ordered on the basis of law, which prevented citizens from doing whatever they wished. Due to this, the Ottoman administration had a positive effect on all peoples living under it, regardless of whether they converted to Islam or not. The Ottoman Empire had an inclusive policy regarding different ethnic groups. This was the total opposite of the policy of the Habsburg Monarchy and the Republic of Venice, where Slavs had a secondary role and could not rise high in the state rulership apparatus. This meant that there were many non-Turkish high-level functionaries in Ottoman state administration and the military. Members of other ethnic groups could rise very high in the Ottoman Empire, but for the highest positions,

---

[7]Strong attack. This attack was carried out by Akinjis, who were heralds of invasion, set loose on neighbouring territory to raid and pillage.

[8](*Akıncı*) Lightly armed irregular Ottoman cavalry.

[9]Another kind of Turkish irregular soldier. The term stems from the Greek word for "armed men". They were members of the Ottoman auxiliary military units in Southeast Europe.

[10]Frontier, i.e. border area of the Ottoman Empire. This was not an officially denoted area with a specific purpose (like the Habsburg Military Frontier), rather a general term.

conversion to Islam was required (Muhaj 2019).[11]

In areas that became part of the empire, there was usually an improvement in the lives of peasants—the most numerous class of society—which led to the acceptance of the Ottoman rule with little resistance. The Ottoman state protected its peasantry because the taxes (*haraç*) that they paid made up a large portion of the state's income. In fact, the protection of peasants from predation by feudal lords was a fundamental principle of the Ottoman state administration (Inalcik 2002). Paying taxes freed peasants from any obligation of military service or conscription, so they were free to focus wholly on agriculture. Thanks to this, peasants were always present on their holdings, as opposed to peasants in Christian European countries, who were regularly conscripted and forced to leave their holdings for months at a time (or even longer). Moreover, as peasants were generally poorly equipped and trained for war, they tended to die in great numbers on the battlefield when conscripted. Lower level aristocrats, European herder groups (Vlachs), and nomadic herder tribes from Asia Minor were all incorporated into the Ottoman military organisation. This gave raiding and pillaging—as legal and common sources of income in the Middle Ages—a new boost. As a result of these factors, new and important political vassals were shaped and included in the Ottoman military organisation by the end of the fourteenth century.

Ottoman leadership had a *wise policy of making alliances* with individual Christian lords and pro-Ottoman political movements within Christian states in Southeast Europe. This is how they dealt fatal blows to enemy states that were already politically disintegrating due to feudal anarchy. They also cleverly used political tensions and rivalries between neighbouring enemy states. Worsening economic and living conditions for the lower classes (in European states) also helped in this regard. All these factors reached a critical point and caused the peasantry, the largest class of society which made up nine-tenths of the population, to stop supporting their feudal lords in resisting the Ottoman expansion. There were even instances of local peasants helping the Ottomans when they were conquering the areas in which they lived, which happened when Ilok, Osijek, Požega, Valpovo, Velika, as well as numerous other cities, were taken (Rogić 1990; Budak 2007).

The significantly better status of peasants under the Ottoman state caused serfs from neighbouring countries to flee into Ottoman territory from neighbouring European Christian states (Matuz 1992). A lack of strong central government in Christian states in Southeast Europe, which would have been able to organise strong resistance to the superior Ottoman military, emboldened local lords in border areas to accept the Ottoman leadership in exchange for retaining their holdings and status. The conflict between the Orthodox and Catholic churches, stemming from the schism in the eleventh century, significantly helped with the integration of the Orthodox Church and population into the Ottoman Empire—as they were fiercely opposed to Catholic ecclesiastic leadership. The Ottoman policy of officially recognising the Orthodox Church, and its patriarchy in Istanbul, along with locally suppressing Catholic Church organisations in conquered areas, allowed them to quickly spread through Southeast Europe (Rogić 1990; Inalcik 2002). In order to avoid battle and enable easier expansion of the state, Ottoman leaders recognised certain individual political units as vassals, which were able to retain a level of self-governance (Wallachia, Moldavia, Transylvania, Basra, the Republic of Ragusa (Dubrovnik), the Crimean Khanate, Kurdistan, etc.) (Matuz 1992).

Because of the aforementioned superior characteristics in relation to neighbouring European states, the Ottomans continually spread their state organisation and were not overly

[11]The position of grand vizier is a good example. From 1320 to 1922, there were 221 grand viziers, of which 118 were not ethnic Turks: 37 Albanians; 31 Slavs (of Bulgarian, Croatian, Russian, and Serbian descent); 11 Georgians; 10 Greeks; 9 Abazins; 4 Arabs; 4 Circassians; 3 Armenians; 3 Italians; 1 Chechen; and 1 Hungarian. The ethnic origin of four grand viziers remains unclear, and it is only certain that these four weren't Turks. It is suspected that two of them were either Slavs or Greeks, while the other two were most likely of either Albanian, Greek, or Slavic origin.

Fig. 6.2 Maximum expansion of the Ottoman Empire in the area of Southeast Europe in the sixteenth and seventeenth centuries (*Source* Regan 2003)

hindered by the occasional loss on the battlefield, like the battle at the walls of Belgrade in 1456. Geographical expansion (into Southeast Europe) was oriented toward territories along the Danube River in the north and the Bosnian state in the northwest (Fig. 6.2).

The Ottomans added newly conquered territories to their administrative–political organisation (Fig. 6.3) and divided them into eyalets and sanjaks. The Ottoman territory extended in a northwest–southeast direction, according to the dominant directions of expansion and how they were able to "break off" territories of neighbouring states that would then join the Ottoman Empire. Additionally, this territorial shape was in keeping with the best-organised caravan routes and the monumental wood and stone bridges spanning the rivers that linked the largest urban settlements (Rogić 1990).

6.3 Croatian Lands Under Ottoman Rule

The first incursions by Akinjis into Croatian territory (in Slavonia and Srijem) took place in the late fourteenth century (Mažuran 1998), while the first pillaging raid in the hinterland of Šibenik was documented in 1415 (Vrandečić 2007). Raiding and pillaging intensified starting in the mid-fifteenth century, especially after the conquest and annexation of the Bosnian Kingdom (1463) and Herzegovina (1482). Right around the fall of Bosnia, there was a marked increase in raids and pillaging of Croatian territory—both of the Hungarian–Croatian Kingdom and the Venetian Republic. The result of these raids was the destruction of the cultural landscape (pastures, cities, and fortifications) and the depopulation of

Fig. 6.3 Ottoman administrative–territorial organisation in the seventeenth century: Eyalets of Southeast Europe (*Source* Regan 2003)

Croatian areas that bordered on Ottoman territory. Documents show how the bishop of Krbava, along with the local population, moved away from the Ottoman border toward the more secure Modruš in the middle of the fifteenth century (1460), while the priest Juraj Cetinjanin led the population of Vrlika in the border area of *Cetinska krajina* (Cetina Frontier)[12] to the northern Dalmatian island of Olib. After Akinjis had devastated their holdings, Kaptol (controlled by the Catholic Church) in Zagreb gave their serfs and peasants a four-year easement of service and tithes in the hope that they would not abandon the land that they worked (Goldstein 2008). By the middle of the fifteenth century, half of the land in Slavonia was uninhabited, due to increasing population exodus (Budak 2007; Rogić 1990).

Venetian sources show that thousands of people in Dalmatia were killed or enslaved during raiding from 1499 to 1539: 10,421 from the Zadar area; 1,689 from the Šibenik area; 309 from the Trogir area; and 215 from the Split

[12]The term *krajina* in this case has the contextual meaning of frontier/borderlands. The name is derived from the Habsburg and Venetian military frontiers with the Ottoman Empire, but has since fallen into common use as a name for specific regions in the Dalmatian Hinterland, e.g. Kninska krajina, Drniška krajina, Imotska krajina, or Vrgorska krajina.

area. Simultaneously there were massive losses in livestock: 90,263 heads of cattle from the Zadar area; 24,837 from the Šibenik area; 6,240 from the Split area; and 400 from the Trogir area.

During wars in the sixteenth century (1501 and 1570), the population of the surroundings of Šibenik relocated to the nearby islands (Vrandečić 2007). Written sources of the powerful Fugger family from 1540 speak of how the surroundings of Zagreb and even the area between Zagreb and Čakovec were abandoned and emptied of the population (Štefanec 2012). In the first half of the sixteenth century, Lika was largely abandoned, as was the Velebit Littoral where the population had deserted its settlements (Rogić 1958). Conversely, Međimurje's population increased steadily as increasing numbers of people from Slavonia and Bosnia flowed in. After the fall of Kostajnica (1556), Nikola Zrinski deliberately resettled serfs from his holdings between the Kupa and Una Rivers (which was then in the border zone with the Ottoman Empire) to Međimurje and western Hungary (Laci 1982).

Thereby, starting in the fifteenth century, and especially during the sixteenth century, population exodus increased and the cultural landscape in Croatian-populated areas (Lika, parts of central and northern Dalmatia, Kordun, Banovina, and Slavonia) was almost completely destroyed—due to the destructive raiding of Akinjis. Apart from the direct consequences of wartime destruction, there were also many indirect consequences. Namely, due to increasing insecurity, foreign merchants avoided living in Slavonian cities. Furthermore, the main trade lines in the Hungarian–Croatian Kingdom, from Pannonia to the Adriatic Sea, had to be moved to the north. Cities that were located on the aforementioned trade routes were gravely affected by this, Zagreb most of all.[13] Moving the bulk of trade to the northern route, through Ptuj and Ljubljana to the northern Adriatic harbours of Bakar, Rijeka, and Trieste, affected cities in the far north, such as Nedelišće and Varaždin (Štefanec 2012).

In the era of the Ottoman Empire's greatest expansion, at the turn of the sixteenth century to the seventeenth century, half of contemporary Croatia was within its borders (Fig. 6.4). Around 300,000 people lived in this area at the beginning of the seventeenth century, and in the Pannonian area, the urban population was double that of Dinaric areas (Moačanin 2000).

6.3.1 Territorial Organisation of Ottoman Croatia

The largest part of the Croatian area under the Ottoman rule was, at first, a part of the great Eyalet of Rumelia, and later of the Eyalet of Bosnia which consisted of the entirety of contemporary Bosnia and Herzegovina. The Eyalet of Bosnia was formed in 1580 as a division of the Eyalet of Rumelia due to the need to more effectively oppose the Habsburgs. Sarajevo[14] was named the political centre of the Eyalet and would grow to become its largest urban centre.

The organisation of the Eyalet into sanjaks, the most firmly directed administrative–political units, was extremely important. Individual sanjaks of the Eyalet of Bosnia covered parts of contemporary Croatian territory (Fig. 6.5). The Sanjak of Herzegovina, for example, with its seat in Mostar, encompassed the greater part of central Dalmatia (apart from the islands) and there was pronounced settlement in the Dalmatian Hinterland (Imotski area). The Sanjak of Klis, founded in 1537, had its seat in Klis despite the fact that its administrator resided in Livno—deep in the interior where it was more secure. The territory of the Sanjak extended from the Sanjak of Herzegovina to the Krka River (however, the islands and the greater part of the coast were in the hands of the Republic of Venice). The Sanjak of Bosnia (with its seat in Sarajevo) encompassed the edges of the Croatian regions of

[13]Up until the sixteenth century, the location of Zagreb on the so-called Senj Route, which led to the northern Adriatic harbour Senj, had been a great boon to the development of the city.

[14]After short stints of administration from Travnik and Banja Luka.

Fig. 6.4 The Croatian area during the zenith of the Ottoman Empire's expansion in the sixteenth and seventeenth centuries (*Source* Dugački and Regan 2018)

Kordun and Banovina. The Sanjak of Krka-Lika encompassed the area of contemporary Dalmatia north of the Krka River (apart from the coast and islands) and Lika, and its political seat was Knin.[15] The Sanjak of Bihać (*Bihke*) was created around 1590, and its seat was Bihać. It included small peripheral areas of the Croatian regions of Kordun and Banovina. The Sanjak of Čazma, founded in 1557, had its original seat in Čazma and changed its name in keeping with moves of the administrative seat first to Pakrac (Sanjak of Pakrac) and finally to Cernik (Sanjak of Cernik (*Cernica*)). It encompassed part of western Slavonia, Moslavina, and the bulk of Banovina.

Although some of the aforementioned sanjaks had seats in Croatian areas, the majority of Croatian territory had a peripheral role within Bosno-Herzegovinian territorial organisation. A characteristic of Croatian areas under the Ottoman rule was that they did not reach the level of development that neighbouring areas in Bosnia and Herzegovina did. The reasons for this can be found in the difficulty the Ottomans encountered in integrating and organising peripheral sanjaks and the strong and persistent guerrilla warfare tactics of Uskoks—which were similar to tactics practiced by Akinjis. After the

[15]The role of seat was temporarily filled by Udbina.

Fig. 6.5 Administrative–territorial division of the Eyalet of Bosnia during the sixteenth and seventeenth centuries (*Source* Regan 2003)

necessary reorientation of the Habsburg and Venetian Military Border, such attacks were intensified and essentially prevented the successful valorisation and integration of peripheral sanjaks, as opposed to the case in Bosnia, which was able to develop without significant interruption. The seats of some sanjaks even had to be relocated deeper into the Ottoman territory, as can be seen in the case of the sanjaks of Klis, Pakrac, and Bihać. Along with the expansion of the Ottoman Empire into neighbouring Hungarian territory at the turn of the sixteenth century to the seventeenth century, two more eyalets were formed that included parts of contemporary Croatian territory. Thereby, the Sanjak of Požega, which encompassed a large part of Slavonia, was given to the Eyalet of Kanije. From 1541 onward, Baranja was a part of the Eyalet of Budim, under the administration of the Sanjak of Mohacs (which had its seat in Pécs). The Sanjak of Srijem (Syrmia) was also formed with Ilok on the Danube as its seat. In the area of the aforementioned sanjaks, Ottoman rulers organised a network of cities in keeping with a new system of administration and territorial organisation (Rogić 1990).

6.4 The Impact of Ottoman Conquests on the Development of the Spatial Concepts of Croatian Regions

After the Ottoman conquest, the spatial definition of the regions of Slavonia and Dalmatia began to change. Slavonia, which had, before the start of the Ottoman raiding, become the new great political core of the Croatian state, was shrunk to its narrow far western belt. The central and eastern parts of the region became constituent parts of the Ottoman state, and the notion that they were part of Slavonia began to fade, but the term was still used for the area of Croatia near the western border of the Ottoman Empire, with Zagreb as its main centre. As Hungarian–Croatian rule was forced from the coastal areas south of Velebit, the use of Croatia and Dalmatia as synonyms—characteristic during the High Middle Ages—also began to fade. Namely, when the Ottomans took the coastal hinterland, the term Dalmatia came to be used exclusively for the coastal areas under the Venetian rule. In relation to this, it is important to also consider the insistence of separation from the concept of Dalmatia (politically) on the part of the Republic of Ragusa (Dubrovnik), by which they sought the support and protection of the Ottoman Empire in times of war, as it was the main adversary of the Republic of Venice in the area.

Until Ottoman territorial regression in the seventeenth century, the geographical concept of Dalmatia consisted of all Adriatic islands (except Lastovo, Mljet, and the Elafiti Archipelago, which were controlled by the Republic of Ragusa (Dubrovnik)), the narrow coastal belt from the conflux of the Zrmanja River to the Cape of Ploča, as well as the areas of the Trogir and Split communes, which included fields of Kaštela and the island of Čiovo.[16] At the same time, under Ottoman rule, the geographical concept of the Lika Region spread to the north and south (in relation to contemporary Lika). With the stabilisation of settlement and the administrative division of the Lika area as an individual sanjak (Sanjak of Lika-Krka), the regional concept of Lika was spatially (re)defined. After more than a century of Ottoman rule, the concept of Lika did not, however, include the area of the Krbava Region[17] which had had a key political and social role in the area prior to the Ottoman era (Pejnović 2009).

6.5 Elements of the New Cultural Landscape

Ceaseless attacks on Croatian territory by Akinjis during the era of Ottoman expansion had a devastating effect on the cultural landscape. Along with population exodus from threatened border areas, elements of Medieval social and economic structures vanished. This process is visible in the abandonment of agriculture, and the disappearance of manor estates and trade fairs. In the cultural landscape of border areas, the well-documented examples of Lika and the Dalmatian Hinterland show how abandoned agricultural land was reclaimed by forests and shrubs (Šarić 2003).

When abandoned border areas fell into the Ottoman hands, their new leaders organised a new type of agriculture that had to have great mobility and flexibility. Due to this, these areas were colonised by an agricultural population which largely practiced transhumant herding. It is clear that herding was safer to practice in insecure areas, as herds could be moved away from the border quickly if the need arose, e.g. in wartime. Because transhumance requires large pastoral areas, these border areas were characterised by a dispersed settlement pattern that did not result in the development of larger settlements.

The majority of the population was concentrated in numerous small, difficult-to-reach settlements and hamlets of the patronymic type.

[16]The southeastern part of the Bay of Kotor and the commune Budva in contemporary Montenegro were also part of Venetian Dalmatia.

[17]Venetian cartographic sources also support this, as there is a clear differentiation of the toponyms Lika (*Licha*) and Krbava (*Corbavia*).

Their irregular forms and structures were a result of spontaneous development, and their location was determined by proximity to pastures (Rogić 1990; Slukan Altić 2003). Planned colonisation of the Muslim (and also Christian) population during the sixteenth century took place as soon as the Ottoman leadership had a firm grasp on newly conquered border areas. The abusive feudal system was replaced by the timar system, which enabled the stronger development of agriculture with an expressed herding component. Strong development of agriculture was spurred by increasing the amount of workable land, construction of irrigation systems, application of modern tools, and especially by the introduction of new agricultural products—rice, corn, tobacco, cotton, mulberries (to cultivate mulberry silkworms for silk production), *Robinia pseudoacacia* (from which exceptionally high-quality honey can be produced—with the help of bees), chestnuts, and new types of fruit and vegetables (Hütteroth 2006).

The Ottoman leadership strictly oversaw land use: the most valuable workable land was given over to crop cultivation while the mountainous areas were designated for herding. Although crop cultivation enjoyed significant advancement, herding continued to be based on transhumant movement and was typically practiced by the Vlach population.

6.5.1 Urban Development—The Oriental-Islamic Town

Urban development proceeded at an exceptionally high level within the organisation of the new network of settlements. The abandonment or destruction of feudal castles raised during the Early or High Middle Ages, along with the collapse of the old social order, represented a break in developmental continuity. New settlements developed in harmony with the new social conditions, as the Ottoman state was able to guarantee peaceful development and the social hierarchy guaranteed security. The security of state territory in the borderlands enabled peaceful political, economic, and social development. Due to this, cities did not have systems of fortifications or even walls that would limit their spatial development (Rogić 1990)

In an economic sense, the Ottoman Empire was self-sufficient in terms of the production of basic goods (agricultural and craft products). The only imports were luxury goods for the needs of the domestic market from Western Europe, the Near and the Far East, (wool and clothing), India (cloth and spices), Russia (fur), and Persia (silk). The economy was based on agriculture—foremost on crop cultivation practiced on state land.[18] (Matuz 1992).

The political power of the Ottoman state was closely linked to its role as the main trade middleman between three continents, i.e. between Africa, Asia, and Europe. Namely, the political core of the Ottoman Empire was in the western part of Anatolia (Asia Minor), which had been part of the great Mongol Empire that stretched from East Asia to the eastern Mediterranean in the thirteenth century. It was the Mongols who linked the two far edges of Asia via imperial caravan routes that made up the so-called "Silk Road". Simultaneously, a second trade route was formed, leading from Arabic countries in Southwest Asia to Eastern and Northern Europe—the so-called "Incense Trade Route". Because of this trade, ships belonging to European traders from Venice, Genoa, the Apennine and Iberian Peninsulas, Flanders, Muscovite duchies, as well as various Mediterranean harbours (including Croatian harbours such as Dubrovnik), plied their wares on the Mediterranean in great numbers.

This made Anatolian cities, particularly Bursa, into centres of exceptionally profitable international trade—among the greatest in the world during this era. Silk, various cloth, dyes, indigo, rice, sugar, soap, medicines, spices, coffee, dried fruit, porcelain, perfumes, gold, silver, and other products were imported from China, India, Persia, Central Asia, and Arabic countries. European merchants sold clothes (mostly made from wool), furs, slaves, iron tools, wine, salt,

[18]Only 5–10% of land was owned privately.

paper, caviar, fish, grain, tin, steel, etc. Over the course of the thirteenth century, the Ottomans would form their state, and go on to conquer and incorporate all major trade centres in Asia Minor —previously controlled by the Byzantine Empire —by the end of the fourteenth century (Inalcik 2002).

The Ottoman Empire was primarily a merchant power and its urban network was adjusted to market needs. All large merchant towns had bezistans (*bedesten*): covered markets that were simultaneously market and warehouse. The Ottoman administration machine systematically maintained caravan routes, bridges, and urban settlements near main roads. Residents of settlements along caravan routes, especially those near mountain passes or canyons and river crossings, were responsible for maintaining transport routes and were thereby freed from compulsory military service and taxation. The existence of new urban centres was linked with numerous innovations in the organisation of caravan routes, upon which rest stops and inns were opened (*caravanserai* and *han*) where merchants and traders stayed—at the state's expense—while travelling. The rapid development of crafts and the concentration of workshops made these locations grow into cities. This was promoted by the state government and had the additional effect of causing Islam itself to spread rapidly, most quickly among the lesser nobility, as well as in rural areas. Furthermore, this encouraged people to settle in urban areas (Busch-Zantner 1938; Rogić 1990; Vresk 2002). The concentration of the Muslim population was the greatest in cities, especially in larger administrative centres, where Muslims often made up more than 70% of the total population.

Ottoman cities had specific spatial, functional, and social structures and were characterised by greater security in relation to cities in neighbouring Christian lands. Examples of the development of the Islamic-Oriental type of urban centre in expansive and stabilised phases of development, when there was no need to build expensive city walls that limited the area of the city in question, are interesting. The results were significantly larger, both demographically and spatially, urban settlements than those in neighbouring Christian lands. Thus, the leading role in the urban network in the Ottoman part of Croatian territory was taken by cities of the Oriental type, and they grew into greater centres than their European counterparts. The highest level of development in this sense was reached by Osijek and Požega[19] (Rogić 1979; Moačanin 2007).

The urban network was hierarchically assembled and cities had different statuses, ranging from the lowest (*palanka*)[20] to mid-ranged (*kasaba*)[21] up to proper cities (*şehir*).[22] The fact that there was no consolidation of a unique urban class within individual cities, largely due to ethnic and religious segregation, is worth noting. This displays the legacy of Byzantine influence in the greater area. The typical type of segregation meant that individual ethnic and religious groups lived together in specific neighbourhoods (*mahala*).

All urban centres were located in the interior, primarily due to the fact that the bulk of the coastal strip was in the hands of the Republic of Venice or the Republic of Ragusa (Dubrovnik). Parts of contemporary Croatia that were located (during the time of Ottoman rule) far from border sanjaks, such as central and eastern Slavonia, western Srijem, and Baranja, enjoyed a high level of development. The agrarian landscape was shaped to a high degree, and large urban centres of the Oriental type played a key role (Rogić 1976, 1990).

[19]Although Osijek, along with Budapest and Pécs, was among the larger cities (*şehir*), the settlement of central significance for the area was Požega, as it was the administrative seat of the Sanjak.

[20]Small provincial settlement. The original meaning of the word is a small wood fort and guardhouse, built to protect travellers travelling through the Ottoman Empire.

[21]Derived from the Arabic word *qaşaba*. A term used for small- or mid-sized urban settlements in areas under the Ottoman rule. To be considered a kasaba, a settlement had to have a mosque, a weekly trade fair, and be oriented toward trade.

[22]Large city.

6.6 Croatian Borderlands

In the early phases (of conquest), Ottoman Akinjis systematically and completely ravaged agricultural land possessed by Dalmatian communes, up to the walls of the cities themselves (Jelaska 1985). However, for reasons relating to trade, Ottoman leaders adopted a cautious policy toward Dalmatian cities, even during times of open hostility and war (Raukar 2007). Thereby, they gave Dubrovnik preferential status as the main merchant harbour for Ottoman products—the fact the Dubrovnik was under Ottoman protection played into this as well. The Venetian harbours Split, Šibenik, and Zadar also played important roles in trade. Many valuable products such as grain, livestock and related products, and wax were exported through these harbours, and salt, individual food products (such as rice), expensive textiles, and finished products were imported (Rogić 1990; Moačanin 2007).

The Ottoman conquest of Bosnia in 1462 was what brought them into direct contact with the territory of the Republic of Ragusa (Dubrovnik), thanks to direct trade links which were made. Due to frequent political tensions between the Ottoman Empire and the Republic of Venice, which grew into open military conflict from time to time, the Ottomans opened trade with Florence. The link between Dubrovnik and Ancona, a free harbour within the Papal States, as well as trade relations between Dubrovnik and the entire Apennine Peninsula—especially with the Duchy of Florence, which would become the Grand Duchy of Tuscany—were extremely important.

Although the Republic of Ragusa (Dubrovnik) paid a yearly tribute to the Ottoman Empire in exchange for keeping its autonomy, Dubrovnik's merchants and traders paid lower duties (2%) than their counterparts from Venice or other European countries (4–5%). Its vassal status enabled Dubrovnik to expand its mercantile reach to the entire Ottoman Empire. Dubrovnik's ships transported wheat, wax, leather, raw silk, and silk clothing from the Ottoman Empire to European states. The import of wool from European countries spurred the development of the textile trade in Dubrovnik, as well as the resulting export of wool clothing to Ottoman cities.

Enclaves of merchants from Dubrovnik popped up in important trade centres such as Sofia, Belgrade, Sarajevo, Edirne, Istanbul, and Bursa. The mercantile success of Dubrovnik also fueled military conflict between the Ottomans and the Venetians during the fourteenth and fifteenth centuries and during two major wars in the sixteenth century—leaving Dubrovnik sandwiched between two vast powers. Despite this, during the sixteenth century, ships from Dubrovnik began to transport goods from distant areas under the Ottoman rule (Egypt, Syria, and Greece) for sale on German and English markets. Starting in the middle of the seventeenth century, there was also a yearly grand trade caravan to Istanbul (Inalcik 2002).

Split was the main Venetian harbour on the Adriatic, which had become an important centre for trade with the Ottoman empire thanks to the fact that its hinterland was under Ottoman rule (Novak 1961). As trade increased in strength, so too did the merchant and tradesman classes, which controlled considerable wealth. Although the array of trade in coastal cities significantly increased, its structure remained unchanged in relation to trade in the High Middle Ages: exports still largely consisted of raw materials for craftsmen such as wood, wool, leather, honey, wax, and furs.

Conversely, border areas of the Ottoman Empire, within the scope of border sanjaks, were harried by constant armed conflict between Ottoman Akinjis and Venetian and Habsburg military units (Uskoks). Due to the constant fighting and destruction, the cultural landscape in these areas was not well developed.

The Neretva Delta and the Makarska Littoral (formally part of the Sanjak of Herzegovina) and the Velebit Littoral (part of the Habsburg Monarchy) were in a unique situation. These areas nominally belonged to the two aforementioned states, but were "no man's lands" in reality, where frequent raiding by Uskoks and Hajduks took place. Due to the general lack of safety and firm government in these areas, they were characterised by a weak settlement pattern, poorly developed existing settlements, and the most poorly developed cultural landscape in Croatian territory.

The long-lasting Venetian–Ottoman conflicts left a deep mark on Venetian Dalmatia, and not just on the coast—the islands suffered as well. Despite the fact that the first war between the Republic of Venice and the Ottoman Empire (1463–1476) didn't have great consequences for Dalmatia itself, in later conflicts (from 1499 to 1502), the Ottoman army razed everything up to the walls of coastal cities. This happened again during the Great Turkish War (Wars of the Holy League) from 1527 to 1540, when the Ottomans took Klis, and again during the War of Cyprus (1570–1573), when settlements on Hvar (Hvar, Stari Grad, and Vrboska) were sacked and burned. By the end of the War of Cyprus, the territorial losses of the Dalmatian communes were great: the district of Zadar was reduced from 1300 to 600 km^2; Šibenik lost half of its territory, reducing it to 570 km^2; Trogir also lost half of its original 250 km^2; and Split lost 10 km^2 of its total 150 km^2. This coincided with a significant drop in population[23] (Rogić 1990; Vrandečić 2007).

Even during peaceful periods, the Akinjis continued their destructive raiding and pillaging. As a typical example, we can take the area of northern Dalmatia, which was sparsely populated because its former inhabitants had migrated to the safer northern Dalmatian islands as a result of constant raiding and pillaging. The consequence of this was the development of the Mediterranean cultural landscape on the northern Dalmatian islands, in conditions of high population density and despite natural conditions that did not favour the development of crop cultivation. In contrast, large complexes of the most fertile soil in the interior of northern Dalmatia that had better natural conditions for agriculture, primarily in the area of Ravni Kotari, became a predominantly herding zone with a poor level of development and a "retrograde" cultural landscape. Records of increases in the price of plots of land indicate the intensity of agriculture and the stronger development of herding on the islands.

Intensification of herding and overgrazing eventually led to the total degradation of land cover and topsoil on individual northern Dalmatian (Kulušić 2001) and central Dalmatian islands (Rogić 1961, 1972, 1990; Carter 1992) under Venetian rule.

In an effort to use agricultural land in the most rational manner, under conditions of autarchic agriculture, land parcels on the islands were divided into four categories: fields, vineyards, pastures, and forests. Arable land was, with respect to its position, further divided into three sub-categories (field, hillside, and hill), and grains (primarily wheat), fruit, and vegetables were sown. Vineyards had the greatest commercial significance and were divided into parcels in fields and on hills. Pastures were of very poor quality and served for raising sheep and goats. Forests were small, and cutting down trees for wood for fuel was strictly controlled.

Insecurity in the coastal parts of Croatian territory, which were ravaged by raids, led to the expressed fortification of urban and rural settlements, and even individual houses. This was most prevalent in the interior of Istria, on the Dalmatian coast, and also on some islands (Šolta, Hvar). Because of this, fortified elements became one of the most dominant and widespread characteristics of the cultural landscape. A typical example of this transformation of the cultural landscape is the area around the Bay of Kaštela, between two important Dalmatian communes: Split and Trogir. Namely, a multitude of small forts popped up[24] where the coast met the fields of Kaštela—which also happened to be the most valuable agricultural land available in the

[23]Zadar fell from 8,100 (the number of inhabitants ten years before the War of Cyprus) to 7,161 inhabitants (1575), Šibenik from 8,220 to 5,310, Hvar from 8,000 to around 5,000, and Korčula from 8,000 to 2,087. In the town Krk in 1575, there were 157 residents, 247 on the island Pag, 1,619 on the island Rab, and the islands Cres and Lošinj together had 5,006.

[24]The toponim *Kaštela* comes from the Latin word *castellum*, which is also the root of the English word "castle".

vicinity. With the construction of 10 castles near Trogir (from the fifteenth century to the seventeenth century) and 11 near Split (from the fifteenth century to the sixteenth century), an acceptable level of security for the local population was achieved. The type of settlement present since the Early Middle Ages underwent a drastic transformation as peasants moved from distant, isolated settlements in the hinterland of the Bay of Kaštela into fortified settlements of a proto-urban type along the coastline. Apart from their function as permanent settlements, castles also served as shelters for the population to hide in during raids (Omašić 1978). Thanks to the efficiency of the system of castles, this type of fortification spread throughout Venetian Dalmatia.

The area of Lika was totally abandoned up to the 1520s. During the phase of stabilisation of Ottoman rule, various Vlach herder groups moved into the area from the Dinaric interior. Due to overgrazing, a significant share of vegetation was destroyed and a landscape dominated by shrubs, primarily in the central part of Lika, had taken shape by the mid-sixteenth century. Due to increasing attacks by Uskoks, Ottoman authorities resettled Muslim farmers from the Eyalet of Bosnia into the area late in the sixteenth century (1576–1586). This led to too much population dispersion. Namely, the Muslim farmer population settled in the areas with the best soil in central Lika, near rebuilt feudal fortifications. This forced the Dinaric herder population to the edges of Lika (Pejnović 1985). Strong development of the Lika Region and the coastline from the Istrian Peninsula to the Konavle Region in the vicinity of Dubrovnik, under the Ottoman rule, was hindered by constant raiding by Uskoks, which originated from the Velebit Littoral (Rogić 1953, 1990).

Opposing tendencies exerted their influence to shape the cultural landscape along the edges of Croatia, which became the core of the most advanced, well-ordered cultural landscapes. A typical example, along with the aforementioned central and eastern Slavonia, western Srijem, and Baranja (under Ottoman rule), would be the area of Istria and the Kvarner–Dalmatian islands under Venetian rule, as well as the territory of the Republic of Ragusa (Dubrovnik) (Rogić 1976, 1990). Simultaneously, in the area of northwestern Croatia, specialisation of traditional peasant farming began to emerge, primarily vineyards (in the Hrvatsko Zagorje, Pokuplje, and Moslavina regions, the surroundings of Samobor, and Kalnik and Bilogora mountains) and pig farming (in Posavina lowlands). These two activities were important sources of income, creating enough wealth that certain peasants became merchants.

This area was also harried by military raiding and pillaging. This was most visible in the crisis of urban settlements, which had developed dynamically in the High Middle Ages. Cities in the area of Croatia north of Gvozd began to stagnate amid increasing Ottoman raiding, decreasing security, and the resulting decreases in economic activity. This resulted in certain cities (e.g. Gradec and Varaždin) losing half of their population over a century (fifteenth century to the sixteenth century). Gradec fell from 3,500 to 1,500–2,000, and Varaždin from 2,000 to ~1,200 (Budak 2007).

6.7 Migration Flows as a Consequence of Ottoman Conquests

The Ottoman era is also associated with significant population changes caused by remarkably dynamic and frequent migrations. The constant expansion of the Ottoman Empire to the north and northwest followed migration flows. Migrations of non-Slavic populations from Asia Minor (e.g. Yörüks) into the area of Croatia and neighbouring countries was negligible, and those that did come assimilated into the local Slavic population over time, especially in places where the masses had largely converted to Islam (Truhelka 1941). Roma people began to settle in the area with the spread of Ottoman rule and they still reside in the area of Croatia and its neighbouring countries to the time of writing.

The main directions of migrations (through Croatian territory) during the Ottoman era were

Fig. 6.6 Migrations during the sixteenth century (*Source* Regan 2003)

southeast to the north, northwest, or west (Fig. 6.6), whereby it is extremely important to differentiate emigration from immigration. Namely, it was mainly the peasant farmer population that was emigrating during the Ottoman era, amid constant attacks by Akinjis, as well as attacks by Uskoks in times of stability or regression of Ottoman rule. Due to this, large groups of the autochthonous Croatian farmer population from Bosnia, Herzegovina, Dalmatia, Lika, Kordun, Banovina, Pokuplje, and Moslavina fled to safer areas like Hrvatsko Zagorje, Međimurje, Žumberak, Istria, the Adriatic islands, Slovenia (White Carniola, Gorizia, and Carniola), Gradišće (Burgenland), Lower Austria, Slovakia, Moravia, Friuli, the Apennine Peninsula (Molise, Apulia, Abruzzo, and the Papal States), and even to Venice and its surrounding territory. Emigration also took place, to a lesser extent, from Slavonia, Baranja, and Srijem, where a large portion of the Croatian population—that continued to live in the area under the Ottoman rule (primarily in Slavonia) during the sixteenth century—had converted to Islam.

The immigration and settlement of farming populations had an expressly positive effect on the development of the aforementioned regions that were final destinations of migrations. We can take Istria as a typical example of an area that was largely abandoned due to frequent Venetian–Habsburg conflicts and epidemics in the border area of the interior, where Habsburg Duchy of

Pazin bordered with Venetian Istria. This caused villages to be abandoned and workable land to become pastures and empty fields, or even to be reclaimed by the forest. Therefore, Venetian authorities began to try to entice people from Croatian border areas who were fleeing from raiding by Akinjis and Uskoks to settle in Istria, primarily by offering various privileges. Habsburg leaders implemented a similar policy in the Duchy of Pazin. Although Akinjis raided the Istrian interior nine times between 1470 and 1499, the Peninsula was far more secure than other Croatian regions. This can be seen in the fact that there were no further Ottoman attacks on the Istrian Peninsula after 1511 (Merlić 2008).

The arrival of people from Dalmatia, Lika, and Bosnia increased the population, and thereby also the income of the Venetian and Habsburg states. Agricultural production was organised into smallholdings (0.2–1.4 ha according to the cadaster of 1571). Productivity was low and there were large amounts of abandoned land in the hands of the aristocracy (Bertoša 2007).

In contrast, in newly conquered areas (usually rather empty in terms of population) under Ottoman rule, herding and farming populations moved in. The nature and direction of these migrations led to the spread of the Štokavian Ijekavian dialect[25] into the Dinarides and also to Mediterranean, peri-Pannonian, and Pannonian areas, primarily by Croatian Catholic herders and farmers from Zahumlje (Herzegovina), and Orthodox Christians from Zeta. Eastern Herzegovinian Štokavian Ijekavian dialect spread into northern Dalmatia, the greater part of Lika, Kordun, Banovina, Slavonia, and Žumberak, and eastern Bosnian Štokavian Ijekavian dialect spread into Banovina, western Slavonia, and Baranja.

The spread of Štokavian Ijekavian was followed by a corresponding decrease in the Čakavian Ikavian dialect to a narrow, disjointed strip of coast and the islands, while Kajkavian Ekavian was forced to the western edge of the peri-Pannonian area. The incidental suppression of Čakavian Ikavian was the most expressed change, as this dialect had dominated the area (the core of the Croatian state) during the Middle Ages: from Istria to the conflux of the Cetina River; all the way to Pokuplje (a region along the Kupa River) in the interior; Posavina (a region along the Sava River); the Dinarides; and the Una Valley. All the islands, apart from those in the far southeastern Adriatic—Mljet, Lastovo, and the Elafiti Archipelago, as well as the western part of the Pelješac Peninsula—also spoke Čakavian Ikavian.

Due to migrations during the Ottoman expansion and the later era of the Ottoman regression, the Štokavian Ikavian dialect spread from Bosnia and western Herzegovina into Dalmatian areas west of the Cetina River and into Lika. It even spread to the far western parts of the islands of Hvar (Sućuraj) and Brač (Sumartin) (Lisac 2003).

An important dimension of the population development of Croatia was systematic enslavement and abduction, which were some of the key elements of raiding by Akinjis and Martolos. Thereby, younger members of the population were particularly valued and would be taken to eastern parts of the Ottoman Empire, most often Asia Minor, where they were sold for work on agricultural holdings or inducted into the ranks of the Janissaries. It should be stressed that there are not enough data regarding the loss of Croatian population to determine exactly how many were taken, and assumptions are based on the calculations of certain authors that are reckoned to be relatively reliable. The calculations range from one-third of the total population (Mažuran 1998) to several hundred thousand people over two centuries, or even up to one million people over the course of three centuries (Kurelac 2000).

[25]The primary dialects are named after the most common question word for *what*: Štokavian (*štokavski*) uses the pronoun *što*, Čakavian (*čakavski*) uses *ča*, and Kajkavian (*kajkavski*) *kaj*. Another common distinction among the dialects is made through the reflex of the long proto-Slavic vowel *jat* and thus the dialects are divided into Ikavian, Ekavian, and Ijekavian, with the reflects *jat* being pronounced /i/, /e/, and /ije/ or /je/, respectively. This can be easily shown using the example of the word "milk", which is *mliko* in Ikavian, *mleko* in Ekavian, and *mlijeko* in Ijekavian.

6.7.1 Vlachs/Wallachians/Morlachs—the People and the Territory

During the era of the Ottoman expansion, one social element, which had played a peripheral role in Medieval society in Southeast Europe until that point, became increasingly important. The element in question was the Vlachs, who were first mentioned in historical sources by Byzantine authors during the tenth century. These authors made a clear distinction between the Vlach and Slavic populations (Mirdita 2004). The first mention of their presence in Croatian areas stems from the fourteenth and fifteenth centuries, and they are referred to by different names: most often as Vlachs, Morlachs, or Ćići. In chronological terms, the first Vlachs arrived in the lands of Croatian feudal lords, fleeting the Ottoman advance into Bosnia in the first half of the fourteenth century. They first came to Cetinska Krajina and the hinterlands of Dalmatian cities (Šibenik, Split, and Trogir), followed by Lika as well as individual Adriatic islands (including the northernmost islands such as Krk and Rab), and the interior of Istria (Rogić 1961, 1975a, b; Mažuranić 1908).

Vlachs are mentioned as a socio-economic group, different from autochthonous Slavic (Croatian) peoples, that exclusively practiced herding. The possession of livestock determined the material status of each Vlach community, which was called a *katun*.[26] Katuns were made up of 50–100 families, and led by a *katunar* (headman or leader). Some katunars came to be the leaders of multiple katuns; these were referred to as dukes. Population growth, as well as an increase in the export of livestock products to the Apennine Peninsula, spurred deforestation and increases in pastoral land. Vlachs probably settled in the Dalmatian Hinterland under Croatian state rule (of the bans Pavao and Mladen II Šubić, and King Ludovik), probably due to the need for soldiers and new settlers to stabilise abandoned areas, and also to increase income from herding (Ančić 1987). Croatian nobles bestowed special privileges upon Vlachs when they settled in the area.

Regarding the question of Vlach ethnogenesis, there is no consensus among researchers (Mirdita 2009). One of the reasons is the fact that the concept of who the Vlachs were/are has often been conflated with other herder groups that practiced the dominant transhumant type of herding during the High Middle Ages, and mainly settled in the Dinaric hilly/mountainous areas of Lika and the Dalmatian Hinterland. Conversely, there were even cases where certain groups of people in the Ottoman Empire, foremost in the Eyalet of Bosnia, were categorised in tax documentation as Vlachs—this was a bureaucratic occurrence, based on the taxes which these groups paid (Moačanin 1998).

The inconsistencies in the understanding of who the Vlachs were—the term was simultaneously used in ethnic and socio-economic senses—stem from the Middle Ages and present a problem for scientific analysis. Namely, Slavic peoples of Southeast Europe referred to the different groups of autochthonous Illyrian–Roman peoples (who were already living in the Dinarides when Slavs arrived in the area) as Vlachs. Therefore it is questionable, and unlikely, that the Vlachs mentioned as living in Croatian areas prior to the Ottoman expansion have any relation to those mentioned in the areas of contemporary Greece, North Macedonia, Bulgaria, Serbia, and Albania. Until the start of the Ottoman expansion, Vlachs who lived within Slavic states in Southeast Europe were either partially or totally slavicised—depending on the state or ecclesiastical organisation that controlled the area—and had converted to Catholicism (Croatia and Bosnia) or Orthodox Christianity (Serbia, Macedonia, and Zeta). Some Vlachs also converted to Islam during the era of Ottoman rule (Truhelka 1941; Goldstein 2008).

Modern scientific research shows how the Vlachs also differed from their neighbours in Byzantine-ruled Southeast Europe in terms of political status, apart from the aforementioned cultural, ethnic, linguistic, social, and economic

[26]Apart from having the meaning of settlement, the word has its roots in the term for mountain shelters used by herders.

differences. In keeping with this theme, their privileged political status in the southern-Slavic area was also present under the Ottoman rule and was one of the main elements that differentiated them from other peoples. Under the Ottoman rule, Vlachs had the right to political autonomy, reduced taxation, and to retain their traditional social organisation into katuns in exchange for taking part in territorial defense and military actions and protecting trade caravans and highways (Mirdita 2004). These measures ensured the loyalty of the Vlachs. Much of the land settled by Vlachs became part of the Ottoman administrative structure and was ruled by various sanjak administrators. This meant that the Vlachs continued to have more autonomy in relation to local aristocrats and timariots. Vlach chiefs had the same social status as the Ottoman military officer class and were given timars to oversee in the same manner (Buzov 2003).

Military actions and conquest in Southeast Europe during the reign of Sultan Murat I (1319–1389) brought the Ottoman Empire into contact with Vlach groups who lived in the Rhodope and Sharr (*Šar*) mountains. The nomadic herder lifestyle of the Vlachs, of which the main characteristic was mobility, meant that they knew the area where they lived very well, making them a perfect fit for the Ottoman conquest machine. Due to this, Vlachs were heavily recruited for the military, despite the fact that some Vlachs still fought against the Ottomans at the beginning of the fifteenth century. By the mid-fifteenth century (around the time that Mehmet the Conqueror took Constantinople in 1453), however, they had warmed up to Ottoman rule, because it guaranteed them privileged social status, autonomy, and self-government—the so-called "Vlach Status". This status bestowed privileges that were not enjoyed by any other peoples ruled by the Ottoman Empire. Therefore, the Vlachs quickly became the cutting edge of Ottoman armies and the main element of Ottoman colonial policy in conquered territory that had been emptied of its previous inhabitants by military action (Mirdita 2004). Akinjis consisting of Vlachs, with the task of constantly attacking neighbouring foreign territory, looting, and destroying all elements of the cultural landscape, would become the new vanguard of the Ottoman army. Because of this, the systematic and planned colonisation of Vlachs took place in step with the expansion of the Ottoman state.

Starting in the mid-fourteenth century, Vlach migrations happened in three main directions, starting from the area of the Rhodope and Sharr mountains. The first led toward Herzegovina and western Bosnia (Livanjsko and Duvanjsko poljes), the second toward central and northwestern Bosnia (Pounje), and the third toward the Drina and Bosna river valleys and onward to Srijem and Slavonia in the north. The strengthening development of crop cultivation and feudal relations in the northern parts of Greece led to Vlach migrations because it increased pressure on all workable land. The increasingly powerful feudal lords were not inclined to allow the best workable land to be used as pastures, so the Vlachs were driven from them and began to migrate north to places where they could practice their herding lifestyle unhindered. The Vlach population ended up settling/colonising far to the north, in the border of sanjaks of the Eyalet of Bosnia—in the area of contemporary Herzegovina and northwestern Bosnia—which they had helped empty and conquer as Akinjis.

Under the Ottoman rule, the Vlach population was settled according to plan into newly conquered areas—areas taken from the Hungarian–Croatian Kingdom and the Republic of Venice—in keeping with the expansion policy of the Ottoman state. This process started in the first half of the fifteenth century in the interior of central and northern Dalmatia (Ravni Kotari, Bukovica, Kninska Krajina, Drniška Krajina, and the surroundings of Skradin) (Mirdita 2009). Further the Ottoman expansion brought planned Vlach colonisation into the area of Lika, Gorski Kotar, Kordun, Banovina, and western Slavonia (Rogić 1958, 1961, 1972, Mažuranić 1908).

References

Ančić M (1987) Gopodarski aspekti stočarstva cetinskog komitata u XIV st. Acta historico-oeconomica Iugoslaviae 14:69–98

Bertoša M (2007) Istarsko rano novovjekovlje: razvojne smjernice od 16. do 18. st. In: Budak N (ed) Dalmacija, Dubrovnik i Istra u ranome novom vijeku. Zagreb, Leykam international, pp 80–120

Budak N (2007) Hrvatska i Slavonija u ranome novom vijeku. Leykam international, Zagreb

Busch-Zantner R (1938) Agrarverfassung, Gessellschaft und Siedlung in Südosteuropa: Unter bes. Berücks. d. Türkenzeit, Harrassowitz, Leipzig

Buzov S (2003) Vlaška sela, pašnjaci i čifluci: krajolik osmanlijskog prigraničja u šesnaestom i sedamnaestom stoljeću. In: Cambi N (ed) Triplex Confinium (1500–1800): ekohistorija: zbornik radova, Split – Zagreb, 2000

Carter FW (1992) Agriculture on Hvar during Venetian occupation: a study in historical geography. Geografski glasnik 54:45–62

Dugački V, Regan K (eds) (2018) Hrvatski povijesni atlas. Leksikografski zavod Miroslav Krleža, Zagreb

Foretić D (1980) Povijest Dubrovnika do 1808., drugi dio – od 1526. do 1808. Nakladni zavod Matice hrvatske, Zagreb

Goldstein I (2008) Hrvatska povijest. Europapress holding, Zagreb

Hütteroth W-D (2006) Ecology of the Ottoman lands. In: Faroqhi SN (ed) The Cambridge history of Turkey, vol 3. The latter Ottoman empire 1603–1839. Cambridge University Press, Cambridge, pp 18–43

Inalcik H (2002) Osmansko Carstvo: klasično doba 1300.–1600. Srednja Europa, Zagreb

Jelaska J (1985) Splitsko polje za turskih vremena. Logos, Split

Kulušić S (2001) Naseljenost i vrednovanje Kornatskih otoka od drevnih vremena do danas (nastavak). Hrvatski geografski glasnik 63:87–120

Kurelac M (2000) Hrvatska i središnja Europa u doba renesanse i reformacije. In: Hercigonja E (ed) Hrvatska i Europa – kultura, znanost i umjetnost: svezak II. Srednji vijek i renesansa (XIII–XVI. stoljeće). Školska knjiga, Zagreb, pp 39–60

Laci S (1982) Razvoj naseljenosti Međimurja. Geografski glasnik 44:51–68

Lisac J (2003) Hrvatski dijalekti. In: Golub I (ed) Hrvatska i Europa – kultura, znanost i umjetnost: svezak III. Barok i prosvjetiteljstvo (XVII–XVIII. stoljeće). Školska knjiga, Zagreb, pp 451–459

Matuz J (1992) Osmansko Carstvo. Školska knjiga, Zagreb

Mažuran I (1998) Hrvati i Osmansko Carstvo. Golden marketing, Zagreb

Mažuranić V (1908) Prinosi za hrvatski pravno-povjestni rječnik. JAZU, Zagreb

Merlić E (2008) Buzeština kroz povijest. Vlastita naklada, Pazin – Buzet

Mirdita Z (2004) Vlasi u historiografiji. Hrvatski institut za povijest, Zagreb

Mirdita Z (2009) Vlasi – starobalkanski narod (od povijesne pojave do danas). Hrvatski institut za povijest, Zagreb

Moačanin N (1998) Introductory essay on an understanding of the Triplex-frontier area: preliminary turkologic research. In: Roksandić D (ed) Microhistory of the Triplex Confinium: international project conference papers, Budapest, March 1997. Institute on Southeastern Europe—Central European University, Budapest, p 125

Moačanin N (2000) Hrvatska i Osmansko Carstvo. In: Hercigonja E (ed) Hrvatska i Europa – kultura, znanost i umjetnost: svezak II. Srednji vijek i renesansa (XIII–XVI. stoljeće). Školska knjiga, Zagreb, pp 63–82

Moačanin N (2007) Hrvati pod vlašću Osmanskoga Carstva do razdoblja reformi u Bosni i Hercegovini (1463.–1831.). In: Budak N (ed) Hrvatsko-slavonska Vojna krajina i Hrvati pod vlašću Osmanskoga Carstva u ranome novom vijeku. Zagreb, Leykam international, pp 108–175

Muhaj A (2019) The origins of the Rivalry between the Ottomans and Venice in the Adriatic. In: Öcalan HB, Karaaslan YZ (eds) Uluslararası Yıldırım Bayezid Sempozyumu, Bursa, November 2015. Türk Tarih Kurumu, Ankara, p 201

Novak G (1961) Povijest Splita: knjiga druga – od 1420. do 1797. god. Matica hrvatska, Split

Omašić V (1978) Topografija Kaštelanskog polja. Čakavski sabor, Split

Pejnović D (1985) Srednja Lika – socijalnogeografska transformacija, Centar za kulturu – Muzej Like, Gospić

Pejnović D (2009) Geografska osnova identiteta Like. In: Holjevac Ž (ed) Identitet Like: korijeni i razvitak. Institut društvenih znanosti Ivo Pilar – Područni centar Gospić, pp 45–84

Raukar T (2007) Studije o Dalmaciji u srednjem vijeku: odabrane studije. Književni krug, Split

Regan K (ed) (2003) Hrvatski povijesni atlas. Leksikografski zavod Miroslav Krleža, Zagreb

Rogić V (1953) Senj: prilog poznavanju položaja i regionalne funkcije. Geografski glasnik 14(15):47–64

Rogić V (1958) Velebitska primorska padina – prilog poznavanju evolucije krškog pejzaža (nastavak). Geografski glasnik 20:53–110

Rogić V (1961) Krk: osobine i postanak današnjeg pejzaža. Geografski glasnik 23:67–101

Rogić V (1972) Primjer geografske analize strukture naseljenosti gorsko-planinskih regija:Ličko goranski prostor. In: Panov M (ed) Zbornik na Jugoslovenskiot simpozium za problemite na selskite naselbi i zemjodelskoto proizvodstvo, Ohrid, May 1972. Sojuz na geografskite instiucii na SFRJ – Geografsko društvo na SR Makedonija, Skopje, p 96

Rogić V (1975a) Osnove historijsko-geografskog razvoja. In: Cvitanović A (ed) Geografija SR Hrvatske – knjiga V: Sjeverno hrvatsko primorje. Školska knjiga, Zagreb, pp 42–45

Rogić V (1975b) Istra: historijsko geografski razvoj. In: Cvitanović A (ed) Geografija SR Hrvatske – knjiga V: Sjeverno hrvatsko primorje. Školska knjiga, Zagreb, pp 184–187

Rogić V (1976) Socio-geografski aspekt dinarskog krša, dinarskog kulturnog areala i dinarskog brdsko-planinskog prostora. Geografski glasnik 38(1):253–269

Rogić V (1979) Slavonska Požega u urbanoj polarizaciji Hrvatske. Geografski glasnik 41(42):71–84

Rogić V (1990) Regionalna geografija Jugoslavije, knjiga 1 – prirodna osnova i historijska geografija. Školska knjiga, Zagreb

Slukan Altić M (2003) Komparativna analiza kulturnog pejsaža ruralnih naselja mletačke i habsburške krajine. In: Cambi N (ed) Triplex Confinium (1500–1800): ekohistorija: zbornik radova, Split – Zagreb, 2000

Šabanović H (1959) Bosanski pašaluk: postanak i upravna podjela. Naučno društvo NR Bosne i Hercegovine, Sarajevo

Šarić M (2003) Turska osvajanja i eko-sistemske tranzicije u Lici i Krbavi na prijelazu iz kasnog srednjeg vijeka u rani novi vijek (15.–16. st.). In: Cambi N (ed) Triplex Confinium (1500–1800): ekohistorija: zbornik radova, Split – Zagreb, 2000

Štefanec N (2012) Grad na prvoj crti obrane. In: Goldstein I, Goldstein S (eds) Povijest grada Zagreba – knjiga 1. Od prethistorije do 1918. Novi liber, Zagreb, pp 108–153

Truhelka Ć (1941) Studije o podrijetlu: etnološka razmatranja iz Bosne i Hercegovine. Matica hrvatska, Zagreb

Vrandečić J (2007) Regio maritima: Dalmacija i Dubrovnik u ranome novom vijeku (1420.–1797.). In: Budak N (ed) Dalmacija, Dubrovnik i Istra u ranome novom vijeku. Zagreb, Leykam international, pp 7–79

Vresk M (2002) Razvoj urbanih sistema u svijetu: geografski pregled. Školska knjiga, Zagreb

7 The Early Modern Era—Ottoman Decline

Abstract

The decreased rate of expansion of the Ottoman Empire on the eve of the Modern Era was also reflected in the internal degradation of the Empire's societal and political systems and coincided with the advancement of Christian states (in societal, political, and economic terms). The Ottoman withdrawal from the Pannonian Basin and parts of Dalmatia was one of the most important events for Croatia during this period, as was the stabilisation of a new border after a series of peace treaties between the Ottoman Empire and the Habsburg Monarchy and the Republic of Venice in the eighteenth century. In the deserted and economically destroyed borderland areas, a policy of settlement and economic valorisation was implemented, largely in Slavonia, the Dalmatian Hinterland, and within the new administrative entity along the border: the Military Frontier. Along with a policy of planned settlement in Slavonia, there were also policies intended to stabilise settlement patterns and the division of land in both Slavonia and the Dalmatian Hinterland. This considerably changed the cultural landscape and enabled an agricultural revolution in commercial grain cultivation. Newly acquired Pannonian territories and economic surpluses created the basis for a new combined transportation axis from Pannonia to the northern Adriatic. This was one of the leading international trade and transport routes during the eighteenth century, until the laying of the first railway. It had great significance for the area of Croatia, but also for neighbouring countries—Bosnia and Herzegovina (under the Ottoman rule) and Serbia—which had previously been linked to international trade by southern Adriatic harbours (Dubrovnik and Split).

Keywords

High Middle Ages · Croatia · Ottoman Empire · Military Frontier · Habsburg Monarchy · Republic of Venice

B. Fuerst-Bjeliš and N. Glamuzina, *The Historical Geography of Croatia*,
Historical Geography and Geosciences, https://doi.org/10.1007/978-3-030-68433-4_7

The Brod Fortress: part of the fortification system of the Military Frontier in detail, Slavonski Brod, Slavonia
The Brod Fortress was built during the eighteenth century (1715–1780) as a part of the Habsburg defense system. The border with the Ottoman Empire approaches to the Sava River, and main trade routes were controlled via important strategic points. It is a monumental example of baroque fortification architecture, rectangular in shape with a star-shaped system of bastions and moats (Photograph: Siniša Sakoman. Reproduced with permission)

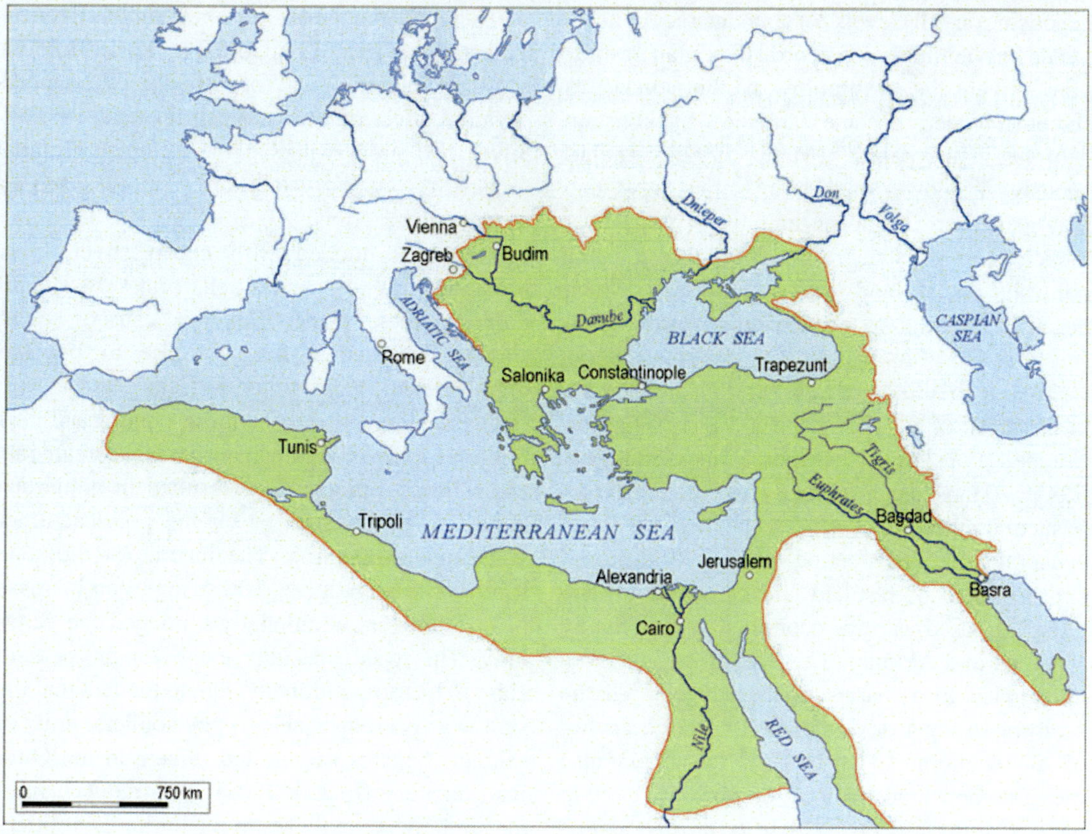

Fig. 7.1 Territory of the Ottoman Empire at the end of the seventeenth century (*Source* Regan 2003)

The transition from the High Middle Ages into the Modern Era was characterised by slower expansion on the part of the Ottoman Empire. This was already evident in the second half of the sixteenth century (Malta, Szigetvár, and Persia), and the major turning point came at the beginning of the seventeenth century (Fig. 7.1) with the end of the Long Turkish War[1] (1606). The peace treaty that was signed at the end of the War stipulated that the Ottoman Empire would gain control of parts of northwestern Bosnia and the Croatian border regions (the eastern part of Kordun and the southern part of Banovina).

These gains were balanced by another stipulation that the Ottomans would cede the region of Moslavina and the newly built border city Petrinja to Croatia, making for a humble overall territorial gain (at best). The decreased rate of expansion was also reflected in the internal degradation of the societal and political systems in the Ottoman Empire, as well as in the advancement of Christian states (in societal, political, and economic terms).

[1]Also called the Thirteen Years' War or Fifteen Years' War, depending on if one reckons the Turkish campaign to capture Bihać to be part of the war or not.

7.1 Internal Degeneration of the Ottoman State

The internal degradation of the Ottoman Empire was a consequence of many interdependent and inseparable factors. The main institutions of the Ottoman state were unable to keep pace with the changes that were wracking the European continent. Increasing corruption, abuses of power, and nepotism on all levels of bureaucracy undermined the authority of central Ottoman institutions. Even the institution of the sultan, as supreme leader, became tarnished by constant debacles in the palace[2] (Matuz 1992). Simultaneously, the power of the Janissaries grew, and they gained an increasing amount of sway over the creation of state policy. As a result, the political, social, and economic crises, which began in the mid-sixteenth century, grew more severe as the years passed.

First, it should be stressed that unending wars in the areas of Southwest Asia (Persia), Eastern Europe (Poland and Russia), and Southeast Europe (Habsburg Monarchy and Republic of Venice) drained the Ottoman state financially and economically. The decline or halt of territorial expansion also resulted in a crisis of the pillaging-based wartime economy (carried out by Martolos and Akinjis). The loss of income from raiding led to a market crisis and exposed the need for stronger development of other branches of the economy. The lack of expansion also meant no new lands for new timars to give out to battered army officers, which led to dissatisfaction among the officer class. Chopping existing timars up into smaller pieces was the solution that emerged, which caused the income of timariots to fall drastically (Busch-Zantner 1938; Rogić 1990; Matuz 1992).

The policy of continuous expansion of state territory, enacted by the Ottoman leadership, required the creation and maintenance of a huge professional army (Janissaries), financed from the state treasury.[3] The formation of a massive army also meant reduced effectiveness of individual fighting units, which could be seen starting in the mid-seventeenth century (Matuz 1992).

The appearance of a huge amount of silver on the European market, imported by Spain from its holdings in the New World, had a pronounced negative effect on the Ottoman financial system. Silver had great value in the Ottoman Empire because the majority of taxes were collected in forged silver coins (*akçe)*. After the Ottoman market was flooded with cheap silver from Mexico, currency values fell, prices rose, and inflation took place. Inflation caused social classes that had a fixed income to become impoverished (e.g. timariots, Janissaries, and state bureaucrats). Under these conditions, corruption increased and spread among bureaucrats and military officers. This resulted in numerous abuses of power such as buying political functions and profiteering. The income of the state itself was also reduced, which led to an increase in taxation that additionally burdened the economy. This was especially tough for the peasant class to bear. Additionally, due to the fact that the state was no longer able to pay soldiers, units of sekban[4] began to roam and pillage in the Ottoman territory (Inalcik 2002). This reduced morale and the overall battle-readiness of military units.

Simultaneously, due to the sudden surge in the economic development of European countries, the price of agricultural products (grains and wool) and raw metals (copper) on the international market rose. The Ottomans tried to control the situation by banning the import of the aforementioned products, which led to an explosion of smuggling. Increases in demand led to shortages of raw materials in the Ottoman Empire, which also pushed the increasing inflation.

As a consequence of the deepening social crisis, there were an increasing number of uprisings involving dissatisfied citizens starting at the end of the sixteenth century. These uprisings primarily took place in the Asian parts of the Empire, especially in Asia Minor—the Ottoman political core (Matuz 1992). Matters also came to a head in Ottoman-controlled Croatia, resulting

[2]This chaotic situation is best shown by the fact that the position of grand vizier (the right hand of the sultan) was occupied by 44 different people over a period of 52 years (1604–1656) (Matuz 1992).

[3]Following the reign of Suleiman the Magnificent (1520–1566), the number of Janissaries grew from 16,000 to 37,000 (1607), while the number of sipahi—mounted soldiers who financed themselves—dropped from around 87,000 to 49,000 (1609); eventually, their number would drop to a mere 8,000 (1630).

[4]Mercenaries of peasant background in the Ottoman Empire.

in the so-called Uprising of Christian *Raya* in 1607 in the Sanjak of Požega. The cause of this was the worsening position of the Christian population amid rising taxes and increasing religious intolerance.

Due to problems related to reduced income during the sixteenth century, the timar system fell into crisis amid illegal annexations, privatisations of state land, and infringements of the right to collect the income. This process is visible in the sudden growth in the number of landowners who began to amass wealth and power on the basis of increased income from agricultural holdings, which was a result of increased tax pressure on the peasant population. This meant that state agricultural lands were transformed into private holdings—called chiflik (*çiftlik*)—on which peasants had to fulfil certain obligations to the owner: the chiflik-sipahi (*çiftlik-sipâhi).* In individual regions, Janissaries, whose influence had grown in relation to the weakening central government, also took possession of timars and turned the local peasants into their servants. It is characteristic that the privatisation of the timar system happened earliest in the Eyalet of Bosnia. Due to this, the degeneration of the system itself was most expressed in the Eyalets of Bosnia and Budim (Busch-Zantner 1938; Rogić 1990).

Under such conditions, a new system of privatised and hereditary land ownership emerged: odžakluk-timars.[5] This system was linked to the concept of Oriental-urban residence. Namely, the process of privatisation of land resulted in the formation of a new land-owning class of timar-sipahis, who lived in cities and exploited the peasantry parasitically by wringing every last bit of income from the land for themselves. This new class monopolised all land income and quickly became very powerful. As a result of the new situation, a part of the peasantry converted to Islam in order to secure a better position for themselves within the socio-legal structure, while another share simply left agriculture and moved to cities where they joined the increasing number of violent gangs that terrorised the country.

Increasing population growth in the sixteenth and seventeenth centuries, and the need for the central government to increase tax income, demanded more agricultural land and stronger agricultural development. Therefore, pastoral areas were reduced and the Vlach population was forced to migrate toward the edges of the Empire. This was closely linked to the limitation of special privileges that the Vlach herder population had enjoyed leading up to this point, and was a source of great discontent.

Concurrently, an increasing exodus of the rural population and abandoned land was noted. This exodus and the corresponding reduction in households that were paying taxes (*haraç*) resulted in conflicts among the local feudal lords and pillaging of the peasant population. Due to the aforementioned problems, in the seventeenth century, the Ottoman social, economic, and political systems fell into a deep crisis (Busch-Zantner 1938; Rogić 1990; Slukan Altić 2006).

The social crisis deepened and weakened religious tolerance—constant war with Christian (primarily Catholic) countries also contributed to this decline. Status, ethnic, and religious segregation in Islamic-Oriental cities prevented the development of an urban class (e.g. *bourgeoisie*), which would have been the bearer of economic advancement and social reform as it was in the West (Rogić 1990). The increase in religious fanaticism in high religious circles, which had been strengthening throughout the first half of the sixteenth century, significantly hindered (or even prevented) the application of scientific and technical knowledge and innovation that was taking place in Christian Europe. Furthermore, these groups used their considerable social and political influence to prevent the further development of science, which they considered occidentalism and against the teachings of Islam. The traditional conservative nature of the Ottoman government also contributed to the lack of scientific development, as the general opinion was that each new innovation could potentially push society into disorder and lawlessness, and also cause losses in state income (Inalcik 2002).

[5]The Eyalet of Bosnia had, at the beginning of the sixteenth century, already accepted the unwritten rule that sons of male members of prominent sipahi families inherited land, which was in opposition to the law.

The Ottoman authorities did not understand, nor were they in a position to find an answer to, the economic changes in Europe. They remained linked to the traditional notion of the state from the Middle Ages. Additionally, they did not tend to the economic development of their state, rather depended on taxation and customs to fill the treasury. New mercantile policies of European economic powers had a destructive influence on the Ottoman free market policy, due to the privileged position of European merchants. Namely, the Ottoman authorities allowed uncontrolled import of goods because the income from customs was the most important source of income for the treasury. This kind of policy, based on exploitation, rather than economic development, was cunningly taken advantage of by European traders. Starting in the mid-sixteenth century, European traders oversaw the entirety of trade between the Empire and the rest of the world. Incentivising imports and limiting exports, primarily in the case of shortages of individual products on the domestic market, led to the stagnation of individual sectors of domestic production (Inalcik 2002).

7.2 New Prosperity in Christian Europe

The internal degradation of the Ottoman Empire was accompanied by a corresponding phase of advancement, in all aspects, in Christian Europe. The sudden trade development of European countries was a consequence of the discovery of new maritime trade routes, and it rendered the middleman role that the Ottoman Empire had played in terms of trade between Europe, Africa, South, Southeast and East Asia much less important. Trade with newly discovered (for Europeans) parts of the world also enabled Atlantic European states to amass great wealth. The development of cities in Christian Europe shaped a new social class—*bourgeoisie* or *Bürger*—that became the bearer of numerous economic innovations in production (manufacturing) and financial business (banks and stock exchanges).

The adoption of new agricultural cultures from the New World, of which corn, tobacco, and potatoes had the greatest commercial significance, spurred a new agricultural revolution. This, along with improvements in nutrition for the European populace, led to the creation of market surpluses and the strengthening of agricultural production.

Political changes were expressed in the destruction of feudal anarchy and the creation of large, strongly directed states. The development of absolutism and parliamentarism, as opposing political tendencies, was accompanied by increasingly complex legislative systems, which guaranteed protection from abuse by authorities to citizens (of certain states). Absolutism significantly aided the rise of certain leading European states, such as France, the Habsburg Monarchy, and Russia, which would become great political and military powers.

The adoption of religious tolerance and gradual liberalisation of society enabled dynamic cultural development, primarily in the states of Western, Northern, and Central Europe. The development of Protestantism laid the foundation for the development of the modern work ethic, on which the advancement of certain European nations (Holland, Denmark, England, Sweden, and Germanic states), in terms of political and economic development, would come to be based. The separation of church and state and gradual secularisation of society freed the minds of the (educated) people and led to the development of rationalism, which would become the basis for the development of modern science and technical–technological revolutions (e.g. the Industrial Revolution) as well as the rapid development of the economies of Christian countries.

7.3 The Military Frontier—A New Territorial Entity

The creation of the Military Frontier is an important historical–geographical phenomenon that took place in Croatian territory, as well as in other Habsburg territories that bordered the Ottoman Empire. At the beginning of the

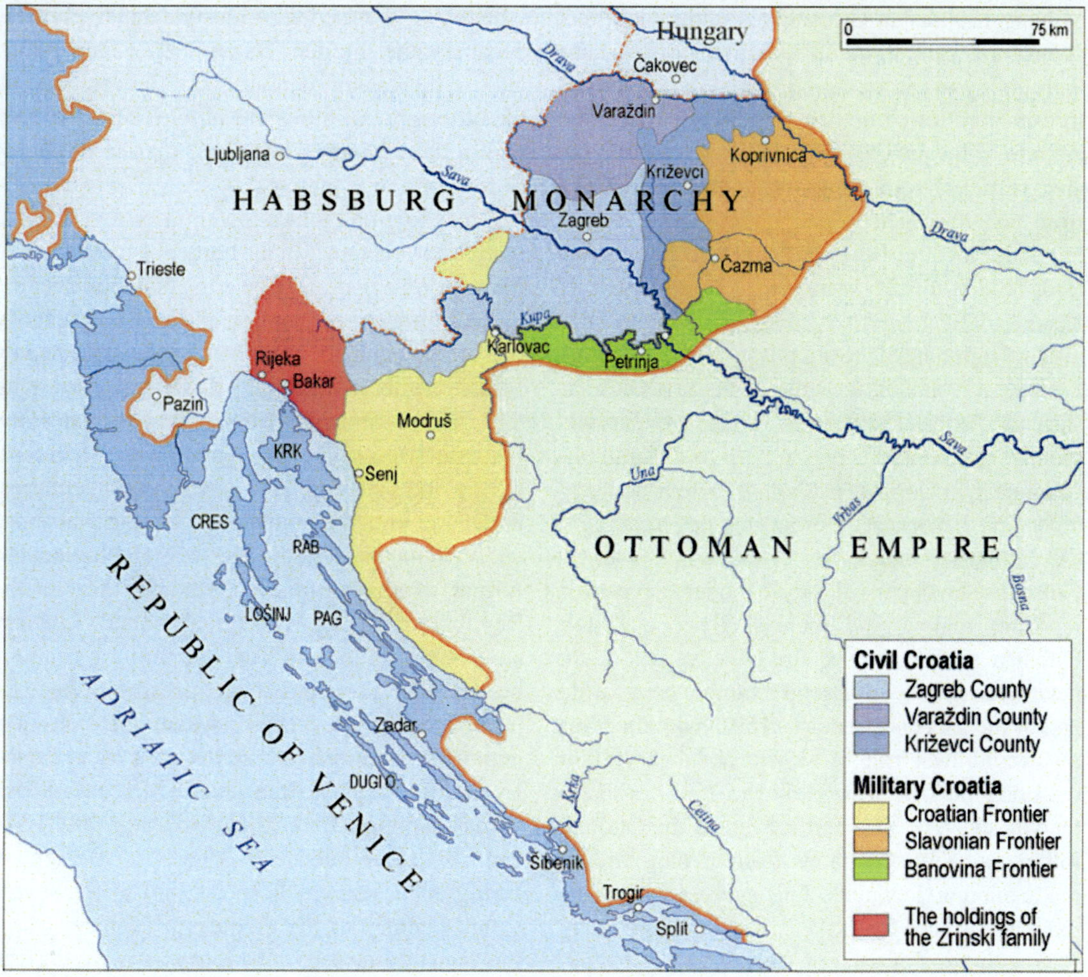

Fig. 7.2 The civil and military administrative–territorial division of Croatian territory at the beginning of the seventeenth century (*Source* Dugački and Regan 2018; Modified after Glamuzina and Fuerst-Bjeliš 2015)

sixteenth century, the border between the Ottoman Empire and the Habsburg Monarchy and the Republic of Venice became largely fixed. The Military Frontier was organised in order to unite the existing piecemeal defenses in a difficult-to-defend region (Fig. 7.2).

The first signs of the emergence of the Military Frontier trace back to the fifteenth century, when the Hungarian–Croatian kings Sigismund of Luxembourg and Mattias Corvinus constructed a defense line along the border with the Ottoman Empire (Jajačka Banovina, Srebrenička Banovina and Senjska Kapetanija).

The original formation of the Military Frontier, and the planned settlement of Vlach groups into deserted and economically destroyed areas of Croatia along the border with the Ottoman Empire, began in 1522. At this point, with the realisation that they didn't have enough strength to resist the Ottoman expansion, the Croatian nobility turned to the Habsburgs for aid. The official succession of the Habsburg dynasty to the Croatian throne was in 1527, and they immediately began to direct the formation of the Military Frontier as a permanently organised border area. It is important to stress how the Habsburg

Military Frontier in Croatia was under the direct control of central authorities, which indicates their policy of discouraging feudal particularism. In exchange for compulsory military service and defense of the area, the Vlachs who settled in the area received some autonomy and were freed from taxation.[6] The Vlach population was thus not under local jurisdiction, which local Croatian lords tended to take issue with (Rogić 1991a, b; Kaser 1997).

The beginning of the organisation of the Military Frontier was around the time of the first Ottoman siege of Vienna (1529) while Akinjis were heavily raiding the adjacent regions (Styria, Carinthia, Carniola, and Gorizia). Although there is no documentation that details the orders for the formation of the Military Frontier, its formal origin was the Croatian–Slavonian Military Frontier. Success in repelling raiding and increasingly open direct Ottoman attacks showed the effectiveness of the new defense policy. A good example came at the end of the sixteenth century (1591) with the Ottoman withdrawal from Moslavina and the victory in the Long Turkish War (Holjevac 2007), whereby the Habsburg government prevented the Ottoman military assaults on Vienna from coming through the Pannonian lowlands. This success was of primary importance for the security of the Monarchy.

The internal crisis of the Ottoman Empire went hand in hand with the revocation of the privileges of Akinjis—that is, among local Vlach herders and the Christian population, who constituted the main body from which Akinjis were drawn—which was the source of much discontent. Raiding and pillaging neighbouring territory was an intrinsic part of the Ottoman territorial expansion process and a significant source of income. The Military Frontier proved effective in reducing or even preventing raiding causing the income from raiding and pillaging to dwindle—from which Akinjis and Martolos derived their personal income—resulting in a corresponding reduction in the rate of territorial expansion. Pastures were also converted to cropland at a higher rate. Thus the pastoral resources, upon which Vlachs based their lifestyle and livelihood, became rather limited (Buzov 1992). This in turn led to increasing banditry and lawlessness as many former Akinjis and Martolos began to raid within the Empire itself, rather than in the lands of the Empire's enemies.

The adoption of the *kapetanija* system (an area ruled by a "captain") in the border of sanjaks led to the creation of a profession army that took over the tasks of protection, customs, and transport security—a role which had, until then, been performed by Vlach Akinjis. The growing dissatisfaction among the Vlachs is shown by several revolts against the Ottoman rule (in 1600 and 1611). Therefore, during the sixteenth and seventeenth centuries, under the encouragement of both the Habsburg and Venetian leadership, a process of settlement of former Vlachs Akinjis in the border areas of the Republic of Venice, Croatia, Slavonia, Hungary, and the Habsburg and Venetian military frontiers began. This (re)settlement became especially pronounced after 1620, when the Ottoman authorities reduced the status of Vlachs in the Eyalet of Bosnia to the level of the non-Vlach Christian population (*raya*) (Moačanin 1998). We can also see evidence of Vlach resettlement in cartographic sources from the turn of the seventeenth century to the eighteenth century, where they were described by various toponyms derived from the word Vlach (e.g. *Valachis*) or Morlach (e.g. *Morlacha, Murlacha, Morlacchia, Morlacca, Morlakia,* or *Morlaquie*[7]) (Fuerst-Bjeliš and Zupanc 2007a).

The organisation of the Venetian and Habsburg military frontiers was a precondition for going on the offensive, effectively turning the tables by putting constant pressure on the border areas of the Ottoman Empire. Paramilitary units (Hajduks and Uskoks) applied the same tactics that had been used so effectively by Ottoman Akinjis and Martolos—pillaging raids—to great effect. This meant that the Habsburg and Venetian leadership had, at this point, effectively adopted the *akın* system, which is best shown in

[6]This special status was detailed in the *Statuta Valachorum* of 1630.

[7]Venetian cartographers (Paolo Forlani, Stefano Scolari) had already applied the term *Morlach* in the Velebit Littoral in the second half of the sixteenth century.

military actions in the sanjaks of Lika and Pakrac, as well as in the relocation of the political seats of the sanjaks of Klis and Bihać to more defensible locations deep in the interior.

Although the Ottomans had taken the entirety of Lika at the beginning of the sixteenth century (1527), which had been largely emptied of population, systematic settlement of the Sanjak of Lika did not happen until the second half of the sixteenth century (after 1577). Ottoman efforts to resettle Lika, however, did not produce good results, primarily due to raiding and pillaging by Uskoks from hideouts along the Velebit Littoral (Rogić 1958). Due to this, the planned colonisation of farming populations (primarily Muslims) if there was, in fact, a plan—showed very weak results, but here (at least) the Ottomans managed to successfully retain the only Catholic (Bunjevci) and Orthodox Vlach herders remaining in service to the empire (Fuerst-Bjeliš 1998).

Hajduk and Uskok raiding, as well as increasingly open conflicts between Habsburg and Ottoman forces in Lika (leading up to the Great Turkish War), would lead to the near-total collapse of the societal structure in the area. The general lack of security meant that the bulk of the population was concentrated in the northeastern and southeastern parts of Lika, as far from the border as possible. Udbina, the largest central settlement in Lika and an important military fortification and political centre, was located in this area (Kaser et al. 2003). Additionally, the frequent pillaging raids by Hajduks and Uskoks, originating from Croatian territory under either the Habsburg or Venetian rule, increased insecurity and hindered normal life in the border areas of the sanjaks of Klis and Požega.

7.4 Cultural Landscape Change Along the Military Frontier and in Civil Croatian Territories

In the Military Frontier, Habsburg authorities invested a great deal of effort into organising defense, renovating and modernising fortifications, and building new fortified cities (Fig. 7.3). A typical example of new planning was the fortified city of Karlovac, founded on July 13, 1579. The greater part of the population, however, continued to be concentrated in rural settlements along the main roadways. Such settlements remain best preserved in peri-Pannonian areas of central Slavonia and in other parts of the former Military Frontier.

The cultural landscape within the Military Frontier was different from other parts of Croatia due to the dominance of herding rather than extensive autarchic crop cultivation. The share of farmland remained smaller than the share of pastures until the beginning of the nineteenth century, and crop cultivation was based on polycultural production of grains (oats, rye, wheat, barley, millet, and corn) and, starting in the mid-eighteenth century, potatoes.

The significance of herding can be seen in the fact that the first veterinary laws in the territory of Croatia (1787) were enacted within the Military Frontier. Extensive pastoralism was a largely backward lifestyle but transhumant migratory herders were still present in the area where Lika meets northern Dalmatia. Their movements became actively regulated by both the Habsburg and Venetian governments starting in the second half of the eighteenth century, via various multilateral agreements (Rogić 1957). The vast forest resources of the Military Frontier were subjected to relentless destruction, which Habsburg authorities tried to mitigate with numerous regulations and prohibitions regarding logging and goat herding (Rogić 1958; Holjevac 2007).

Documents that record the increasingly severe deforestation in border areas of Croatia, i.e. the Military Frontier, during the seventeenth century, also show increases in population density (Štefanec 2003). Similar processes were also documented in the interior of Istria under the Habsburg rule (Rogić 1975a). Unplanned logging in the foothills, primarily in mountainous Croatia and the Velebit Littoral, began the process of degradation (Rogić 1958; Holjevac 2003). Due to this, starting in the mid-eighteenth century, these parts of Croatia underwent a process of deagrarisation and corresponding processes of the development of other economic activities (primarily trade, transit, and

Fig. 7.3 The Military Frontier at the end of the eighteenth century (*Source* Dugački and Regan 2018)

woodworking) as well as emigration processes (Rogić 1958; Kaser 2003). Around the same time, Habsburg leadership also began to invest significant effort into halting the process of deforestation and enacting a sensible policy regarding forest upkeep (Gruenfelder 2003).

During the Great Turkish War, Ottoman Lika lost most of its population, and Habsburg leaders made great efforts to resettle it after taking control of the territory. From 1685 to 1686, Bunjevci from Dalmatia and the Velebit Littoral settled in Lika, as well as Croatians from the upper Kupa River area. Smaller Catholic and Orthodox Vlach groups from the Eyalet of Bosnia and Dalmatia (Bukovica) also moved into the area. According to the results of the first Austrian census of Lika (1712), Catholic and Orthodox Vlach families made up 74% of the total population[8] (Kaser et al. 2003).

Gorski Kotar was also largely unpopulated and economically bereft until the 1670s, and significant results were achieved in repopulation starting at the beginning of the seventeenth century (Rogić 1954; Pavić 2012). With the inclusion of the area of Krbava and Lika in the Military Frontier, under the administration of the

[8] It is worth mentioning that the term "Vlach" was not used consistently in censuses. Although the term was mostly used to describe the Orthodox population, it was sometimes also used to describe ethnic affiliation (as we can see with the term "Catholic-Vlach" that was most often used to describe Bunjevci).

Lika Regiment, the geographical concept of Lika expanded to include part of the Military Frontier (Rogić 1977; Pejnović 2009).

The first signs of organised defense in Venetian Dalmatia can be seen during the Cretan War (1645–1669). Its effectiveness, however, was demonstrated in the total destruction of the Ottoman elements in the cultural landscape (cities, forts, rural settlements, farmland, and forests) in the Dalmatian Hinterland. This destruction would have the greatest consequences for the area of contemporary Dalmatia. Dalmatia was, in this era, mostly within the Ottoman Empire, while a narrow coastal belt, from the conflux of the Zrmanja River in northern Dalmatia to the Makarska Littoral (Podstrana) in central Dalmatia, was ruled by the Republic of Venice. Venetian Dalmatia also included all islands apart from Mljet, Lastovo, and the Elafiti Archipelago, which belonged to the Republic of Ragusa (Dubrovnik). The wartime strategy of both sides was based on a "scorched earth" policy and the systematic destruction and pillaging of conquered forts, urban and rural settlements, as well as all agricultural areas under the control of the opposition. Even some individual cultural–historical artifacts, like the Roman amphitheatre in Salona, were destroyed. Northern Dalmatia was hit worst by all of this, as were individual parts of central Dalmatia and the surroundings of Split, where, apart from the larger coastal cities, all elements of the cultural landscape were destroyed. The Cretan War was characterised by horrific violence and anarchy among the numerous irregular brigades made up of members of the local population, who fought for both sides—as it suited them.

The worst consequences of the war were famine and plague epidemics, as well as population regression. Namely, the interior of central Dalmatia and parts of the coast south of Zadar were almost completely depopulated. Within the area controlled by Venice, the greater part of the population abandoned the interior and the coastline and moved to the islands, which were spared the ravages of constant warring. The Venetian army burned all the rural settlements that were abandoned after their former inhabitants moved to the coast. Concurrently, there was an influx of Christian refugees (both Catholic and Orthodox) coming from areas under the Ottoman control. These were people from Bosnia and Herzegovina who the Venetians colonised into the continental areas under their rule, while those coming from the Makarska Littoral were resettled onto the islands (Brač, Hvar, Vis, and Korčula).

Despite immigration, the Cretan War represented a major population regression for Venetian Dalmatia. Such that the three largest Dalmatian cities—Šibenik, Zadar, and Split—had lost half of their population by the end of it (in 1682 Šibenik had 4,171 residents, Zadar 3,597, and Split 3,350). Apart from war damages, famine, and plague, the lack of sufficient trade during the 25-year conflict also contributed to this depopulation (Novak 1944).

The fact that the Venetian Military Frontier was formed later is a consequence of the shape of their territory in Croatia: a narrow strip along the Adriatic coast. The Venetian authorities did not control enough agricultural land to enable them to colonise Vlachs into their territory until the end of the seventeenth century. When the first Vlach populations—formerly in the Ottoman service—were settled according to plan in Venetian Dalmatia (in the surroundings of Šibenik), their presence led to stronger trade for Dalmatian cities. Additionally, planned settlement of Vlach groups, fleeing from Lika and Bosnia, into Venetian Dalmatia, took place during the Cyprian War. It was in this area, abandoned during the War, where the Republic of Venice organised its own Military Frontier along the border with the Ottoman Empire, following new territorial acquisitions and the shift of the Venetian–Ottoman border deep into the interior. The following frontiers were organised: *Kninska Krajina*; *Drniška Krajina*; *Cetinska Krajina*; *Imotska Krajina*; *Vrgoračka Krajina*; and *Neretvanska Krajina*,[9] (Fig. 7.4) where Vlach

[9]The historical Croatian term for frontier is *krajina*. Today, the term *krajina* as a toponym is also used as a regional concept. The aforementioned frontiers—i.e. krajinas formed during the era of the Military Frontier along the Habsburg and Venetian border with the Ottoman Empire—have persisted as regions to present day in parts of the Dalmatian Hinterland.

Fig. 7.4 Croatian territory under the Venetian rule and the area of the Venetian Military Frontier from the fifteenth century to the eighteenth century (*Source* Dugački and Regan 2018)

populations from Lika and Bosnia and Herzegovina, which had been in service to the Ottoman Empire as Akinjis until that point, were (re)settled (Buzov 1993).

The settlement of Vlach populations was carried out in border areas of Croatia (with the Ottoman Empire) under both the Habsburg and Venetian rule. These areas were largely abandoned when the planned settlement began. As a result of continual war and pillaging, nearly all elements of the cultural landscape had been destroyed. This resulted in the use of the Latin term *terra deserta* (abandoned land), used by Croatian geographers of the day to denote the area on their maps; Venetian cartographers used a similar term—*terra distructa* (destroyed land)—for the area. Cartographic depictions of Croatian territory from that era (regardless of which regime the maps were made for) show the destruction of the cultural landscape in the southern and eastern border areas (Rogić 1990; Fuerst-Bjeliš and Zupanc 2007a).

The parts of Croatian territory that were not under the Ottoman control, or within the Habsburg or Venetian military frontiers, showed positive developmental tendencies due to an influx of farming population from the prior era of the Ottoman expansion. The need for effective defense resulted in the revitalisation of fortified cities and the strengthening of central (royal) leadership. The presence of armies in the Military Frontier resulted in the development of agriculture and market growth, largely undertaken by Italian and local merchants. Increases in population and settlement density and the stronger development of crop cultivation were most expressed in Hrvatsko Zagorje, Istria, and on northern Dalmatian islands. However, disorganised settlement was not without problems, as the sharp increase in population made the question of land distribution, especially farmland and pastures, a pressing issue. Despite regulations regarding land distribution, corrupt local administrators often divided land as they saw fit, which resulted in chaos (Merlić 2008).

Thanks to continued trade with the Ottomans, even during times of open war, the economies of Dalmatian cities under the Venetian rule were revived. Habsburg authorities supported the development of the northern Adriatic cities of Senj, Rijeka, and Bakar to serve as their main Adriatic harbours (Rogić 1954, 1956, 1958, 1969a, 1971). The Habsburgs adopted a more mercantile-oriented policy at the beginning of the eighteenth century by declaring free trade on the Adriatic (1717) and naming Rijeka and Trieste as free harbours (1719); Kraljevica took on the role of main military harbour in 1725 (Rogić 1969b; Budak 2007). The development of mercantile harbours was favoured for strategic reasons. Trieste, for example, was far from the Ottoman border and thus more secure (Pavić 2012). It is important to stress how Venetian Croatia was put under great fiscal stress due to the crisis in the Republic of Venice, caused by a shift in the focus of maritime trade from the Mediterranean to Atlantic Europe, as well as the massive cost of constant war with the Ottoman Empire. The suppression of communal self-governance resulted in weaker trade, while a parasitic collection of rent fees put pressure on land workers (Novak 1961).

7.5 Ottoman Territorial Retreat and the Development of a New Border

Stemming from internal political conditions within the Ottoman Empire, individual powerful military circles pushed for military action, which resulted in the Great Turkish War (1683–1699) against the Habsburg Empire. The crushing defeat of the Ottoman army at the walls of Vienna in 1683 marked the start of the Habsburg offensive. In the following years, the Austrian and Polish armies, which had fought in the Pannonian lowlands, joined the Republic of Venice in attacking Dalmatia and the Peloponnesus, while Poland struck throughout Eastern Europe (Moldavia and Podolia).

Concurrently, reinforcements began to arrive from other European states, especially the Holy Roman Empire and France, and Russia began their own war with the Ottomans by attacking Azov in 1687. This had the Ottoman army fighting major engagements on three fronts. Because of their obligation to defend their vassal—the Crimean Khanate—the Ottomans pulled significant military strength from the Pannonian Basin. This allowed the Austrian army to strike quickly and deeply into the region. The Austrian offensive led to the total destruction of the Ottoman cultural landscape in the Pannonian Basin. As the Austrian army moved further into the Pannonian lowlands, the area began to empty as the Muslim population fled to the south. The same situation took place during the short-term attacks by the Austrian army into Bosnia, Serbia, and Macedonia. With the Treaty of Srijemski Karlovci (Karlowitz) in 1699, a large amount of territory changed hands between the Ottoman Empire and the Habsburg Empire, and the border was moved to a line following these rivers: Tisa–Bosut–Sava–Una (Fig. 7.5) (Rogić 1990).

Fig. 7.5 Territorial changes and borders after the Treaty of Srijemski Karlovci (Karlowitz) in 1699 (*Source* Regan 2003)

The borders of the Republic of Venice expanded from the narrow coastal belt deep into the interior of the Dalmatian Hinterland, along with the simultaneous annexation of the Bay of Kotor (*Linea Grimani*). In the area of Croatia, the border shifted from the headwaters of the Zrmanja River to Knin, Vrlika, Sinj, Trilj, Duar (Zadvarje), Vrgorac, and Gabela (on the right bank of the Neretva, upstream of Metković). All of the aforementioned forts were located in settlements that were cut off from the greater part of their natural hinterland, as the border was an hour's walk from each. Each had a strong garrison, which meant that the role of these settlements was primarily a military one, while the development of other economic activities was limited. The Great Turkish War, which broke out 15 years after the end of the Cyprian War, led to the destruction of the cultural landscape in Ravni Kotari and Bukovica in the hinterland of Zadar, the central Dalmatian Hinterland, and the Neretva Delta. By the war's end, the entire area, which had been largely emptied of population, was annexed by the Republic of Venice and called the "new acquisition" (*Acquisto nuovo*) (see also Fig. 7.4) (Novak 1944).

Although the Venetians had taken the greater part of eastern Herzegovina in 1694, cutting the Republic of Ragusa (Dubrovnik) off from the Ottoman Empire by land, the buffers of the Neum and Sutorina corridors were formed by the Treaty of Srijemski Karlovci (Karlowitz) at the request of Dubrovnik. This prevented the Venetians from conquering the Republic of Ragusa (Dubrovnik), which had long been one of the strategic goals of the Venetian leadership. The unsuccessful Habsburg attack on and short occupation of Bosnia, and the spiteful burning of Sarajevo, was largely the result of resistance on the part of the local population. Namely, thanks to the stable population and economy in the Eyalet of Bosnia, the Ottomans were able to mobilise and organise a coherent defense. The attacks on Bosnia and Serbia resulted in the migration of rebellious Christians (who had aided the invading Austrian forces), following the retreat of the Habsburg army, and their resettlement into the Pannonian lowlands.

By the end of the next war, fought between the Republic of Venice and the Ottoman Empire from 1714 to 1718 (the Habsburg Monarchy entered the war in 1716), new territorial gains were obtained with the Treaty of Požarevac (Passarowitz) (1718). The result was the movement of the border between the Habsburg Monarchy and the Ottoman Empire south of the Sava and Danube rivers, by which total strategic control over transit on said rivers was achieved, and the peri-Pannonian part of Serbia was also annexed. In the southern part of Croatia, the contemporary border of Dalmatia was shaped (*Linea Mocenigo*), thanks to the so-called "newest acquisitions" (*Acquisto nuovissimo*) of the Republic of Venice in the Dalmatian Hinterland (see also Fig. 7.4). The formation of the aforementioned Neum and Sutorina corridors during the war was formalised by the Treaty of Požarevac (Passarowitz), thereby shielding Dubrovnik from the ambitions of Venice.

According to the peace treaties, the right of sea access for the Ottoman Empire (in the Eyalet of Bosnia) was confirmed via two narrow corridors named for the small rural settlements of Neum and Sutorina. This formal recognition mostly benefited the Republic of Ragusa (Dubrovnik), which would have become completely surrounded by the Republic of Venice otherwise. In reality, neither of the corridors were connected with their hinterlands nor did they have any necessary preconditions to develop into effective harbours for Ottoman use.

The Ottoman Empire resisted Russia effectively in their war from 1735 to 1739 (the Habsburg Monarchy joined in 1737) and showed its strength by the end of the war (Treaty of Belgrade, 1739) by returning the border to the Danube and Sava rivers.

A new war between the Ottoman Empire and Russia broke out in 1787, and the Habsburg Monarchy joined the war in the same year. The Habsburgs withdrew from the war in 1791 with the signing of the Treaty of Svištov (Sistova)—which confirmed the status quo, along with minor corrections of territory in favour of Croatian territory under the Habsburg rule, specifically in the area of Drežnik and Cetin (in Kordun), that moved the border to the Una River—annexing eastern Lika in the process (Fig. 7.6). Some of the refugees from the Serbian and Bosnian Eyalets, who came to Croatian territory under the Austrian rule during the various wars, were settled into Slavonia and the surroundings of Karlovac. This resulted in the foundation of 12 new settlements in the surroundings of Drežnik and Cetin (Iveljić 2010).

7.5.1 Implications for the Spatial Concept of the Slavonia Region

In keeping with the suppression of the Ottoman rule, the concept of Slavonia began to be used again for parts of the Pannonian Basin between the Sava, Drava, and Danube rivers. The concepts of Upper and Lower Slavonia,[10] which were physically separated by the military territory of the Varaždin Generalate, persisted. With

[10]"Lower" refers to the area of Slavonia in which the Ottoman rule was suppressed.

Fig. 7.6 Territorial changes in Croatia and Slavonia during the seventeenth and eighteenth centuries (*Source* Dugački and Regan 2018)

the re-establishment of the županija structure in Lower Slavonia, the new counties sent their representatives directly to the Hungarian parliament, which led to the further division of Lower Slavonia from Upper Slavonia in a political sense. From this point on, the area of Upper Slavonia west of the Military Frontier was not understood to be part of the region of Slavonia. The additional Hungarian formation of the Kingdom of Slavonia in Lower Slavonia defined the borders of the contemporary region of Slavonia.

7.6 The New Pannonian–Adriatic Developmental Axis and Its Implication for the Cultural Landscape of Slavonia and the Military Frontier

With the inclusion of the Pannonian area into the Habsburg Monarchy, the cultural landscape began to transform. Aspects of the aforementioned changes are still visible in the cultural landscape at the time of writing. Namely, this newly conquered, steppe-forest area had been emptied of the population by years of war. When it arrived, the Habsburg army promptly destroyed all remaining elements of the Ottoman cultural landscape. Just before the end of the Great Turkish War, Slavonia had between 300,000 and 320,000 inhabitants, and the first Habsburg census of Slavonia after the war reported only one-third of that number (Mažuran 1988). Habsburg leaders decided to resolve this problem with a policy of planned settlement of population from Central Europe, as well as with Christian refugees from the Ottoman Empire, primarily into Slavonia and other parts of the Pannonian Basin. The most numerous immigrants were Catholics from areas of Bosnia along the Drina and Sava rivers (Podrinje and Bosnian Posavina regions), who were called *Šokci*.[11]

Along with the policy of planned settlement, there were also policies intended to stabilise settlement patterns and the division of land. This enabled an agricultural revolution in commercial grain cultivation. The economic advancement of this area of Croatia was followed by sharp growth in population in new, planned rural settlements that were laid out in precisely defined square parcels, allowing for the creation of large landholdings. Concurrently (at the turn of the sixteenth century to the seventeenth century), a process of land consolidation also started in the northwestern part of Croatia. This was made possible by a decrease in the Ottoman attacks as well as by new, sensible management policies (Rogić 1953, 1956; Budak 2007). Starting in the 1760s, a planned concentration of settlements was undertaken, as reforms aimed at strengthening crop cultivation and administrative control of the population were not effective on a population spread throughout dispersed settlements. Karst areas, however, were not suitable for such a policy, and the end results were humble at best. The single exception to this was central Lika, along the main roadways and the contact zones of various poljes and Velebit Mountain (Rogić 1970, 1973; Pejnović 1985).

High levels of agricultural production, which the Ottoman authorities were never able to achieve, spurred the development of transport; this rose from the need to export surplus agricultural products—primarily grains—to Central and Western European countries. Work was begun on refurbishing the Pannonian–north Adriatic transportation system of rivers and river canals (Fig. 7.7). The Habsburg state put great effort into river canal works for river transport, and also dug new canals: Bezdan–Bečej and Maroš (Moriš, Maros, Mureş)–Tisa, and built new harbour facilities and systems for towing boats upstream. The main precondition for the development of the entire transport system was the construction of the Danube–Sava–Kupa canal system. New wagon roads were also built that ran from Karlovac, the westernmost river harbour in the system, to Adriatic harbours.

Via Carolina Augusta[12] led to the harbour Bakar and was opened for use in 1736. Via Josephina[13] led to the harbour Senj and was opened in 1799. These two roads, particularly Via Josephina, were built to the highest standards of road building available at the time, making them some of the most modern roads in Southeast Europe. The network of roads was later extended, routing Via Josephina through the Istrian Peninsula from Pazin across Učka Mountain to Kastav, thus linking Istria with Kvarner (1785). Via Theresiana[14] connected Lika and Kvarner through Velebit, from Gospić to Karlobag (1750–1754), and the pre-existing mountain road from Velebit to Dalmatia (Žuta Lokva–Gospić–Knin–Zadar) was also modernised (1786–1789).

Via Lousiana[15] was built at the beginning of the nineteenth century (1803–1811) and linked Karlovac with the harbour Rijeka. A road that

[11]A Croatian ethnic group who were settled into Slavonia, Srijem, Baranja, and Bačka, in the border area between contemporary Croatia, Hungary, and Serbia. Most *Šokci* in Croatia ended up in eastern Slavonia and western Srijem.

[12]Via Carolina Augusta was named after Emperor Karl VI.

[13]Via Josephina was named after Emperor Josef II.

[14]Via Theresiana was named after Maria Theresia.

[15]Via Louisiana (also Via Ludovicea) was named for Maria Luisa, daughter of Emperor Franz Josef I.

Fig. 7.7 The Pannonian–northern Adriatic transportation system

connected the harbours Rijeka and Bakar (the terminal point of Via Carolina Augusta) was also constructed at this point. Two other important roads were also opened: Via Kordun from southern Lika through Kordun to Karlovac (at the end of the eighteenth century) and Via Triestina from Zagreb to Ljubljana and on to Vrhnika, along the Sava River. Many local roads were also rebuilt. Road building works had long-term consequences for the area of Lika, where Gospić was given special stimulus for its role as the new development centre of the region.

Agricultural goods were also transported upstream along the Danube and Drava rivers toward Slovenia and Austria, although this transportation corridor lagged considerably behind the Sava Corridor. River transport was a great boon to the development of harbour areas. Among the cities that profited most from this were Zemun, a city on the right shore of the Danube near the Danube–Sava Conflux (part of contemporary Belgrade); Osijek, a city on the right shore of the Drava, near the Danube–Drava conflux; and Vukovar on the Danube, Varaždin on the Drava, Sisak on the Sava, and Karlovac on the Kupa (the final point for river transport on its way to the coast) (Rogić 1992). Transport of goods from Slavonia to northern Adriatic harbours typically took eight to ten days (Valentić 2003).

Increased trade between the Habsburg Monarchy and the Ottoman Empire additionally contributed to the development of the Pannonian–north Adriatic transportation system. Trade increased due to the treaties of Srijemski Karlovci (Karlowitz) and Belgrade, wherein the easing of Austrian tariffs on all goods imported from the Ottoman Empire was negotiated (Mirdita 2009). Individual Ottoman trade caravans with goods from Bosnia, which had been bound for Dalmatian harbours in the Venetian territory, started reorienting toward the northern Adriatic harbours Senj, Rijeka, and Bakar. During the eighteenth century, mutual trade relations between individual harbours in Kvarner (specifically Bakar, Kraljevica, Rijeka, Lovran, Senj, and Karlobag) and Dalmatian harbours was strengthened, as was trade between Rijeka and Trieste (Peričić 1998; Rogić 1975b).

The building of the roads that linked Karlovac with northern Adriatic harbours was a significant boon to the development of settlements located along them. This was most visible in Gorski Kotar. In that part of Croatia, in the eighteenth century, strong population migration took place, mainly heading from isolated areas toward settlements along roads. Owing to this, Delnice grew into the role of the main centre of the region.[16] The transport of goods along the roads from Karlovac to Adriatic harbours became one of the most important activities in this part of Croatia. People from neighbouring regions came to take part in the trade, i.e. from Carniola and the Kvarner Littoral—even from distant parts of Central Europe like Czechia.

The Pannonian–north Adriatic transportation system was, up until the laying of the first railway, one of the leading international trade and transport routes during the eighteenth century. It had great significance for the area of Croatia, but also for neighbouring countries—Bosnia and Herzegovina (under the Ottoman rule) and Serbia—which had previously been linked to international trade by southern Adriatic harbours: Dubrovnik and Split. Bosnian and Herzegovinian import and export focussed increasingly on the Pannonian–north Adriatic transportation system and would continue to do so even after the emergence of the first railways (Rogić 1975a, 1981).

In the same time period (during the eighteenth century), the Military Frontier underwent a change as the formerly dominant extensive pastoralism was replaced by crop cultivation, under more favourable conditions of settlement and land parcelling. Planned colonisation was followed by further development of rural settlements along roadways and consolidation of landholdings. This type of settlement pattern was a precondition for the development of advanced agricultural production. The regular shape and layout of land, as well as the uniform appearance of houses, was the result of planned parcelling. The transformation of the cultural landscape was a consequence of the fact that crop cultivation had taken over the main role within the total economic structure, largely thanks to administrative acts on the part of Habsburg leadership (such as the acts from 1768 and 1769 regarding the obligation to cultivate potatoes and corn). These changes made the transhumant iifestyle less relevant (Rogić 1990; Pejnović 1985; Slukan Altić 2003).

The exception to this was in the area of southern Lika. The area was partially settled in 1712 and only in rare cases was this settlement planned and organised. Newly settled families took the lands that suited their fancy and raised settlements on their own initiative. Such settlements were dispersed throughout a wide area, and only some were larger settlements made up of a cluster of smaller hamlets. All efforts by Austrian authorities starting in the eighteenth century to group the dispersed settlements into a tighter form were unsuccessful. Such that a cultural landscape of small and dispersed land parcels (2–3 ha on average) was created, and individual holdings of large size were the exception rather than the rule. Individual land parcels were divided into three categories. The most important was called *okućnica* (a small personal garden around the house where traditional peasant polycultures were grown). The second category was called *livada* (meadow) and was used for grazing. This type made up the

[16]This process sped up after the Via Louisiana was built, as it was the best mountain road of the era.

majority of landholdings (more than half). The smallest share of landholdings was the third category, called *oranica* (workable land and fields), consisting of approximately one-tenth of the total landholdings (Blanc 1953; Rogić 1990; Kaser et al. 2003).

Accordingly, planned colonisation did not change the economic structure because the bulk of settlers practiced (sheep) herding. They also began raising cattle, which indicated the stability of settlement. Furthermore, the need for larger pastures and the dominance of herding reinforced the expressly dispersed settlement pattern. The crop cultivation component in agriculture was complimentary. This meant that this type of dispersed holding with concentrated agricultural lands (primarily cooperative or communal pastures) was preserved in Lika. As the population grew over time, these small clusters of farms grew into patronymic settlements (Rogić 1982, 1990; Pejnović 1985).

The most frequently cultivated cultures on holdings in the continental part of Croatia were various grains (wheat, oats, barley, rye, and millet) which increasingly suppressed traditional cultures such as spelt. However, despite the high level of production, additional grain still had to be imported. Wine, on the other hand, was an export product. Vineyards were widespread in all parts of Croatia, apart from mountainous areas. Of all the new cultures, corn spread especially fast, as did potatoes (after the eighteenth century). The cultivation of industrial plants (hemp, linen, tobacco, sugar beet, and to a lesser extent silkworm) also increased. Both older and newer systems of crop rotation were in use; winter and summer crops were rotated and the land was left fallow as needed. Traditional methods of working the land (by hand or with the help of beasts of burden) prevailed, using old (often wooden) tools and plows. Due to this, crop yields were weak, even in fertile years, which meant that many rural families (who grouped together in cooperatives and practiced an autarchic type of agriculture) scraped by on the edge of hunger. Herding in Croatia has always been extensive, with emphasis on cattle, pigs, and horses. Thanks to the rich Slavonian oak forests, pig farming developed heavily, based on acorns. Merchants even bought the right to harvest acorns and brought their pigs from the Eyalets of Bosnia and Serbia (Karger 1963; Iveljić 2010).

The lack of an economic policy that would have incentivised and protected domestic production led to an imbalanced position in relation to more developed parts of the Habsburg Monarchy. The main generators of production (saw mills, sugar mills, and grain mills) first came into being on aristocratic holdings, as the aristocracy based its advancement on trade capital. This sort of processing of raw materials was begun on the initiative of the state government with money from foreign investors. This foreign capital was used to open the first large sugar, potash, liquor, and tobacco manufacturing facilities in Rijeka, which was the largest centre of manufacturing in Croatia from the turn of the eighteenth century to the nineteenth century. Habsburg authorities incentivised the development of silk production in Slavonia, the Military Frontier, and Varaždin. Woodworking and various mills were flourished in Žakalj (near Rijeka), Vukovar, Karlovac, and Zagreb, and there were sugar mills in Virovitica, located in the Drava River lowlands.

The production of potassium carbonate (Gorski Kotar, Slavonia, and Kvarner), leather (Rijeka, Osijek, Karlovac, and Kutjevo), tobacco (Rijeka and Požega), bricks (Nova Gradiška and Vukovar), glass (Mrzle Vodice, Trakošćan, Varaždinske Toplice, Osredek, and Velika), matches (Osijek), beer (Osijek, Karlovac), and paper (Rijeka, Zagreb, and Vidovec) took place as well—but in a somewhat smaller amount. At the same time, in Gorski Kotar (primarily in the area of Čabar), the first small manufacturing facilities based on the use of local raw materials and oriented toward woodworking, ironworking, and glass production were founded.

New elements of the cultural landscape were mines, from which valuable raw materials were extracted for use in manufacturing. Iron was extracted from mines in Gorski Kotar and in the continental part of Croatia on the slopes of Samoborska Gora (Rude), Medvednica (Bliznec Stream Valley), Petrova Gora, and Zrinska Gora mountains. Copper was also mined in the settlement of Rude near Samobor, and sulphur in Hrvatsko Zagorje (Radoboj) (Iveljić 2010).

Shipbuilding, mainly oriented toward building wooden sailing vessels, flourished in the northern Adriatic harbours Rijeka, Bakar, and Kraljevica. These three cities boasted a combined six shipyards and produced nearly half of all Austrian sailing vessels that were intended for long voyages in this era. Shipyards in Senj had lesser significance and fell behind the aforementioned three large shipbuilding centres over time. Shipbuilding of river vessels was a healthy industry in Sisak, the important final point of upstream transport on the Sava River—thanks to the great volume of transport along the Pannonian–north Adriatic transportation system. In larger harbour cities located on the aforementioned river transport system, markets for grains, wood, and tobacco were also built.

There were many problems with manufacturing, such as inadequate markets and competition from imported products from other parts of the Habsburg Monarchy—foremost from Austria itself. Due to this, most production facilities closed within a few years of opening. Despite the high rate of failure, these first facilities provided the initial spark that lit the fires of industrial production in Croatia.

7.7 The Cultural Landscape of Dalmatia—A Reflection of the Venetian Politics

Along with the spread of the Venetian holdings in Dalmatia, the question of its valorisation emerged, especially due to problems arising from the natural and social characteristics of their newly conquered territory. The karst relief is an important natural characteristic to emphasise, as it has very poorly developed soil and meagre vegetation. This means that the area generally offers very poor conditions for the development of agriculture. Social problems stemming from the weak settlement as well as an exceptionally low level of economic development were consequences of wartime abandonment.

7.7.1 The Impact of Land Reform on Migration and the Cultural Landscape of the Venetian Acquisto Nuovo

In newly conquered parts of Dalmatia, the Venetian authorities implemented a new agrarian policy based on exclusive state ownership of land. However, this was not the only system; two concurrent and opposing systems of land ownership were present in the Dalmatian territory. The old system, based on private land ownership, was still valid in areas of the so-called old acquisition (*Acquisto vecchio*), the border of which was called *Linea Nani*. This area had come into the Venetian hands at the expense of the Ottomans before the start of the Great Turkish War. In the newly annexed territory east of the aforementioned border, however, agricultural land could not be sold, divided, or exchanged because it was all owned by the state. In this area, the Venetian government doled out land to the population (or the Church) as tenures via investiture. The land was divided among the tenants living on it, including Vlachs who moved into the area from the Eyalet of Bosnia, according to the number of people per household (Fuerst-Bjeliš 2003; Ćosić 2012).

This was a system that the Venetians had already put into practice on Crete in the thirteenth century, according to which land was divided into land for permanent use and land for hereditary use, with the condition of a decade of state service in peace or service in wartime. The largest holdings were given to worthy aristocrats, in order to secure their loyalty to the Venetian government, as well as to save on administrative costs in newly conquered areas. By applying this system, the newly arrived population settled into the area quickly (Vrandečić 2007).

Apart from settlers from Bosnia and Herzegovina, parts of the Dalmatian Hinterland that had been abandoned during various wars were resettled by people from over-settled parts of Dalmatia, lured by promises of greater privilege. Some of the farming population from the northern Dalmatian islands were resettled into Ravni

Kotari in the hinterland of Zadar, while a portion of the population from the central Dalmatian area of Poljica (around the city of Split) was resettled deep into the interior of Cetinska Krajina. Similar local migrations, predominantly moderate in size, also took place in other parts of Venetian Dalmatia. Although (re)settlement had already begun during the Great Turkish War, migrations from the neighbouring Eyalet of Bosnia would continue through the three wars between the Republic of Venice and the Ottoman Empire that would follow. The settlement was especially strong in 1789, during the last conflict.

Furthermore, there was planned resettlement of skilled workers from the Venetian hinterland (mostly from Friuli) for work in budding factories and on large landholdings. There was also resettlement of craftsmen, bureaucrats, and physicians into coastal cities.

Following the Great Turkish War, which emptied the Dalmatian Hinterland of its population, most of the agricultural land was abandoned. Therefore, the policy of planned settlement and development of settlements became closely linked to agricultural valorisation. This was supported by corresponding ordinances, according to which the cultivation of grains and fruits—primarily olives and grapes—and maintenance of access ways and water sources were mandatory. The Venetian cadastral sources show that the population was constantly increasing and settlements developing throughout the eighteenth century, which indicates the positive results of this settlement policy. The average family in rural areas at the time had eight to ten members. Autarchic crop cultivation, based on polycultural production on very small land parcels (0.35–1 ha), in combination with transhumant herding, prevailed in the economic structure. A cultural landscape of open fields and dispersed landholdings with a dispersed pattern of small, irregular patronymic settlements and hamlets took shape. These were indicators of the limitations of the karst environment and the area's limited amount of workable land. The most valuable land, in karst poljes, was saved for crop cultivation, while herding developed in areas of high karst or in the foothills surrounding karst poljes (Fuerst-Bjeliš 2003).

The Venetian leaders settled the Dalmatian Hinterland primarily out of strategic military reasons, without regard to the consequences that colonisation had for the Mediterranean natural environment and its extremely sensitive ecosystem. Namely, the prominent role of herding led to overgrazing, which, by the end of the eighteenth century, caused degradation of plant cover and soil, as well as the shaping of a "bare karst" landscape in northern Dalmatia (Fuerst-Bjeliš and Zupanc 2007b). This general picture of a rocky, waterless, infertile area was contrasted by travel writers of the era (e.g. Yriarte 1883) with notes regarding locally chosen "green and fertile" areas. This primarily was in regard to the narrow Krka and Cetina valleys and karst poljes. Yriarte described these areas within the region as places in which one felt "the breath of life, in contrast to the wasteland that prevailed" (Fuerst-Bjeliš and Kale 2018).

Via special laws, such as *Legge Grimani* (1756), the Venetian authorities continued to advance and intensify crop production on the best lands in the northern parts of the Dalmatian Hinterland, where the cultivation of specific cultures (linen, hemp, olives, pears, chestnuts, plums, cherries, mulberries, figs, almonds, hazelnuts, and honey) was prescribed by various regulations. A regulation was enacted that stripped the right to work a given piece of land (from whomever was using it) if it lay fallow for two years without interruption. These laws spurred the development of herding and ordered the mandatory cleaning and maintenance of irrigation canals and wells. At the same time, herding—of goats in particular—was limited to mountain areas. These laws, which were very forward-thinking and placed emphasis on improving the lives of peasants, ended up being only partially effective because they were inconsistently enforced. Furthermore, the retrograde form of extensive herding with very limited cultivation of crop cultures—largely used as livestock feed—continued to prevail.

The Dalmatian Hinterland did not enjoy positive development trends like the parts of Croatia under the Habsburg rule did. This was largely due to degradation in hilly areas due to overgrazing, as well as flooding and the conversion of farmland into swamps in lowland areas (Fuerst-Bjeliš et al. 2003).

This was additionally affected by the collapse of the political system of the Republic of Venice, which was riddled with corruption and nepotism. The aforementioned system of ownership relations in agriculture, which was not consistent in all parts of Dalmatia, added to this. The exception to this was the islands, on which private ownership of land dominated, apart from forests and pastures which were considered public goods. The domination of private ownership and institutions enabled the formation of large landholdings worked by colonists and day labourers. There was only a formal difference between colonists and day labourers: colonists possessed their own small landholdings for personal use, while day labourers had not yet managed to obtain a piece of land for themselves. Furthermore, landlords were not able to remove the aforementioned workers (colonists and day labourers) as long as they consistently and properly worked the land and paid their rent. Neither colonists nor day labourers were formally regarded as serfs, making their status nominally better than their fellow peasants in other European countries. This was largely a meaningless distinction, as they didn't live better in reality. They were obligated to give their landlords anywhere from one-fourth to one-half of their crop yield (in addition to the rent they paid), determined by crop culture and the direct assessment of the groundskeeper (*gastald*) employed by the landlord.

In contrast, in the newly acquired parts of Dalmatia (*Acquisto nouvo* and *Acquisto nuovissimo*), the Venetian leaders divided the land via hereditary investiture to worthy servants of the state, who had proven themselves in the war against the Ottomans or in some other suitably impressive fashion. The process of dividing land was accompanied by numerous acts of corruption, leading to the land falling into the hands of those who were not objectively numbered among "the worthy". The new land-owning families could only lose their land if the male line died out, they emigrated, or the land remained unworked for three years in a row. This shaped a new land-owning class, many of whom had never even set foot on the land which they rented to colonists. They dealt with colonists via agents, who were paid a percentage of the amount of rent collected, which often resulted in theft, abuse, and bribery. This parasitic absentia landlord system had a negative impact on worker motivation among the peasants, and reflected negatively on the natural environment due to soil exhaustion (in pursuit of greater yields); poor development conditions and outmoded agricultural techniques and tools only added to the workers' hardship. The best example of this is the underuse of manure fertiliser (despite the large number of livestock) and the preponderance of wooden plows (only at the end of the eighteenth century would iron plows be introduced to the area).

Around the time that the Republic of Venice began to fall apart (1797), the "forest and pasture" category of land was the most common in Dalmatia. This meant that orchards, vineyards, and grasslands made up a significantly smaller share of total land.[17] Moreover, there was more swampland than orchards and vineyards combined. At this point in time, only one-sixth of the land in Dalmatia under the Venetian rule was worked.

Fishing was, as opposed to crop cultivation and herding, less autarchic and most developed on the islands near Šibenik and in central Dalmatia, as well as the Makarska Littoral. Salted fish was the main product of the island Vis, as well as smaller islands like Šolta. Although a large portion of the catch was consumed locally, some were exported to other parts of the state. On the islands Krapanj and Zlarin, near Šibenik, coral and sponge harvesting from the seafloor also developed; the entire catch was routinely shipped to Venice to be processed and sold for high prices. In Venetian Dalmatia, there were

[17]There were 910,800 ha of forests and pastures, 150,000 ha of orchards and vineyards, and only 50,000 ha of grasslands.

two large salt refineries, on the islands Pag and Rab.

The salt refinery on Pag was responsible for the lion's share of production. Although demand for salt on the market, for food and other uses (salting fish and meat, and use in manufacturing), was secure, salt refining was barely profitable. The cause of this can be found in the policy that sent three-fourths of salt produced directly to Venice, leaving only one-fourth for the salt producers to sell on the market. Due to this, Dalmatia only satisfied one-third of its own needs for salt, which was mostly used for preserving fish. Because of Venice's exploitative policy, there was no great interest in the development of new salt refineries (Peričić 1998).

7.7.2 Manufacturing, Transportation, and Commerce

The exploitation of raw materials from mining enjoyed a boom in Dalmatia during the eighteenth century. The most exploited raw materials were coal, iron, and tar. The main incentive was given by the Venetian authorities, who wanted to ensure that the mines were producing enough to satisfy their manufacturing needs. Tar was extracted in the surroundings of Vrgorac in the Dalmatian Hinterland, near the city of Trogir, and on the island of Brač. It was usually transported directly to Venice to be used in shipbuilding. Starting in the mid-eighteenth century, coal began to be mined on the islands of Pag and Brač, in the surroundings of the city of Skradin, and on Promina Mountain in the Dalmatian Hinterland. All of this coal, which brought the mine owners great wealth, was exported to the Habsburg Monarchy (Rijeka and Trieste) for use in manufacturing. Modest iron ore deposits were exploited for a short time in the Dalmatian Hinterland near the cities of Sinj and Vrlika; however, they were quickly exhausted. Furthermore, rocky sand was quarried on the islands Hvar and Vis for glass production in Venice, as well as lead (Pb) in Knin (Peričić 1998).

In order to reduce the import of manufactured goods from other parts of Europe, the Venetian leaders started developing manufacturing capabilities in Dalmatia. However, the results were disappointing because most of the production facilities, even those that used only domestic raw materials, closed within a few years of beginning production. This was the fate of production facilities for leather, textiles (using domestic wool and hemp), bags (made from goat hair), ceramics, and bricks (using local clay). The main problems were the lack of capital and trained workers. New facilities for textile dyeing were built in the Dalmatian Hinterland in the eighteenth century, but these were exclusively for the domestic market as opposed to the existing facilities in coastal cities. These new facilities were characterised by outdated technology and weak production capacity. Based on local raw materials, the first manufacturing facilities making silk and candles (Split, Zadar, and five facilities in Šibenik), distilled rosemary oil (on the island Hvar and Šolta), dried tobacco (Nin), and bricks (Neretva Delta) were established.

The only Dalmatian products from manufacturing facilities that made it to the wider European market were wine, liquors, and iron. Liquor production facilities grew out of traditional distilleries, using locally grown fruit (mostly cherries in this case). The main production centre was Zadar, but facilities were also present in Šibenik and in Mali Lošinj. Liquor production brought a large amount of income because the product was easy to sell both domestically and abroad—it even made its way into some royal liquor cabinets.

Shipbuilding was a traditional trade that enjoyed modest advancement in the eighteenth century. The largest shipyards were located in Trogir and on Korčula, where the fleets of Pelješac and the Bay of Kotor were built. There were also many smaller shipyards, located in Dubrovnik, Split, Zadar, and on the islands Vis (Komiža), Brač (Milna, Supetar), Hvar (Hvar, Vrboska, and Stari Grad), and Lošinj (Mali Lošinj).

The first iron foundry was fueled by high-quality iron ore from mines in the Dalmatian Hinterland (Cetinska Krajina—between Sinj and Vrlika). This foundry was an important incubator for industrial production in Dalmatia and employed several hundred workers (mostly foreigners). The facility mainly produced iron rods and cast blocks, the bulk of which was exported to the Apennine Peninsula (Peričić 1998).

The development of Dalmatian coastal cities was limited by the status of maritime trade (the main traditional occupation), and the development of the hinterland was limited by finite natural resources and weak autarchic agriculture (barley, wheat, sorghum, millet, and various vegetables—especially cruciferous varieties). Namely, trade between Dalmatian harbours and the rest of the world was possible only through Venice, and only on the Venetian merchant ships —as was largely the case in the Middle Ages. In contrast, local initiatives undertaken by individual Dalmatians, aimed at continuing advancement of the economy by founding various institutions (e.g. the Economic Society of Split from 1774), fell through due to the lack of interest on the part of the Venetian authorities. Under such conditions, even the *Legge Grimani* was disregarded in large parts of the region (Novak 1944).

Because Dalmatia was not blessed with enough of its own resources to spur economic development, and the social structure additionally hindered the advancement of the existing economic system, the single prosperous occupation was trade with the neighbouring Eyalet of Bosnia. However, Bosnia was also in a crisis of sorts due to its poor transport connections. The Venetian authorities did not want to maintain the roads that connected Dalmatian harbours with Bosnia due to fears that in doing so they might aid a future Ottoman invading force, in the event that war broke out again. The only well-maintained local roads that linked individual small coastal trade settlements with centres in the interior were: Skradin–Ostrovica; Makarska–Imotski; and Trogir–Drniš (Pederin 2011). It was only near the end of the Venetian rule that some of these old roads (e.g. Šibenik–Skradin) were redone and new roads were built (Šibenik–Knin and Zadar–Knin). Road maintenance was a function of trade growth with Bosnia, allowing Bosnian caravans to more freely carry Bosnian products to Split (the most important harbour and trade centre in Dalmatia), Šibenik, Zadar, Skradin, Makarska, and Metković.

7.7.3 Population

Under conditions of general economic backwardness, population growth presented a problem. During the eighteenth century, the population of Venetian Dalmatia (including the Bay of Kotor, which was a part of the territory at the time) grew from 108,090 (in 1718) to 288,320 (in 1795). This was a growth of over 150% over slightly less than 80 years. No other part of Europe grew as much over the same period.

Only a small percentage (10–12%) of the population of Dalmatia lived in cities. The majority of the population lived in a large number of rural settlements—roughly 800—that were made up of very poorly built small houses. The consequence of population growth under the conditions present in Dalmatia at the time (economic backwardness, limited natural resources, and development level) was abject poverty. This was especially expressed in rural areas and increased the pressure on the region's already strained resource base. The average population density (20 people per km^2), however, was significantly below the European average.

Additionally, in terms of age structure, there was not a strong component of the working-age population. In the second half of the eighteenth century, data show that roughly half of the population belonged to the elderly category or were younger than 16. When one reads individual documents, which note the lack of population in certain Dalmatian areas (particularly in the hinterland), one gets a better picture of the region's weak level of economic development in this era.

Difficult living conditions resulted in emigration, beginning at the start of the eighteenth century. The main migration waves went to the Pannonian Basin (to the Banat, Slavonia, and Srijem regions), to neighbouring Austrian areas of Croatia (Lika), and to the Eyalet of Bosnia. In isolated cases, such as the emigration of 126

Orthodox Christian families from northern Dalmatia in 1758 (due to agitation on the part of the priesthood), emigres made for distant lands like southern Poland or Russia. The largest emigration was documented in the second half of the eighteenth century, primarily from the interior parts of northern and central Dalmatia (Cetinska Krajina) (Peričić 1998).

The report of J. Boldue gives additional information on the social structure of Dalmatia, according to which the region was dominated by rural peasants (93.5% of the population), while the urban population made up a distant 4.4%, nobility a mere 2.2%, and craft and tradesmen the last 0.7%. General retrograde tendencies in society and dependence on meagre agriculture resulted in frequent famines[18] (Borković 1982). Namely, the dietary base of the Dalmatian population depended on a small number of locally produced crop cultures (millet, sorghum, barley, cabbage, and white onions) and complimented them with meat, eggs, milk, fish, and butter. Depending solely on this sort of autarchic type of agriculture could have dire consequences, i.e. famine, during dry summers and cold winters (Peričić 1998). Additional development problems in Dalmatia caused plague epidemics, which spread via trade routes from the neighbouring Eyalet of Bosnia. The disease hit the border areas in the Hinterland hardest, as well as coastal trade cities that engaged in trade with ships from the Ottoman Empire. Large epidemics were documented in the periods of 1730–1734, 1763–1764, and 1783–1784. The disease also spread to the area's livestock herds. By creating a sanitary cordon in the border areas of Dalmatia, the Venetian authorities were able to stem the epidemic(s) somewhat.

This caused additional documented emigration from Dalmatia as famine and plague took their toll. Near the end of this era of hunger and epidemics, the population that had emigrated to Bosnia had largely returned, which was not the case with the population that migrated into the territory of the Habsburg Monarchy (Soldo 2011).

7.8 The Republic of Ragusa (Dubrovnik) and Istria—Developments and Crisis in the Borderlands

Thanks to a long period of peace, the Republic of Ragusa (Dubrovnik) achieved one of the most developed (agri)cultural landscapes on the Croatian Littoral. Due to great efforts on the part of the peasantry, as well as remittances that they received from family members who were sailors, many were able to purchase their own land from members of the nobility. This was the case throughout the Dubrovnik Littoral, Konavle, and on Pelješac Peninsula; despite the fact that peasants were not legally allowed to own land in these parts of the Republic. Despite the fact that Dubrovnik's government had formally banned this practice in the second half of the eighteenth century, it continued, manifesting formally as a leasing arrangement or as a transformation of a peasant into a member of the *bourgeoisie* (who were free to purchase land), after the peasant in question moved to a city and thereby ceased to be a peasant. Although the Republic of Ragusa (Dubrovnik) had not been ravaged by famine, the lives of the peasantry were difficult due to heavy taxation. Namely, the nobility had to gather wealth via taxation after control over maritime trade passed into the hands of the *bourgeoisie*.[19]

The results of the oldest preserved census of the Republic of Ragusa (Dubrovnik) (1673–74), undertaken by clerical leadership, show the bulk of settlement in the regions of Konavle, the Dubrovnik Littoral, and Pelješac Peninsula; each of which had roughly 5,500 inhabitants. The area

[18]The largest famines were documented in the following years: 1715–16; 1731; 1733; 1738; 1740; 1747; 1752; 1757; 1764; 1772–1775; 1777; 1779.–80; 1781–1782; 1783; 1784; and 1793.

[19]This particular economic policy resulted in peasant uprisings in Konavle (1799–1800). The increasing dissatisfaction of the peasantry, as well as the increasing political ambitions of the *bourgeoisie*, was rooted in the passiveness of the population in relation to French authorities dissolving the Republic at the beginning of the nineteenth century (Foretić 1980).

near Dubrovnik itself, known as Astarea, was especially densely populated—likely around 50 inhabitants/km^2—similar to the islands of the Elafiti Archipelago. Areas with more workable soil were significantly less settled: Konavle had 25–30 inhabitants/km^2; the Littoral around 20 inhabitants/km^2; and Pelješac Peninsula roughly 15–20 inhabitants/km^2 (Krivošić 1990). Approximately two-thirds of the total population of the Republic of Ragusa (Dubrovnik) lived in these three moderately populated, agriculturally developed areas (in the second half of the seventeenth century.) These areas were periodically characterised by the emigration of population "surplus", indicating the high agrarian density throughout the Republic. In eras of population growth, high agrarian density quickly became oversettlement.

The political limitations and continual oversight over immigration into Dubrovnik, which was systematically undertaken by Dubrovnik's authorities, show how the Republic was under constant pressure from immigration. Additionally, this helps explain and confirm the recorded levels of agrarian density.

Conversely, sudden growth in the shipbuilding industry in the mid-eighteenth century led to an increasing exodus of the agricultural population into a more prosperous profession: maritime work, i.e. sailing. This put greater taxation pressure on the peasants who remained in the fields. Under such conditions, when there was a lack of population in a given rural area, Dubrovnik's government solved the problem via the planned settlement of population from the hinterland of the Republic, especially from neighbouring Herzegovina.

In this era, maritime capital within the Republic dispersed, and the main shipbuilding centres became Orebić on Pelješac Peninsula, Slano, and Cavtat. The prominence of Pelješac Peninsula, where two-thirds of Dubrovnik's merchant fleets were based (1640–1680), also became more evident (Ćosić and Vekarić 2003). The Republic itself continued to preserve its strong trade component, and Bosnian merchants exported the greater part of their wares through Dubrovnik's various harbours (in the face of increasing competition from harbours in Venetian Dalmatia—Split in particular).

To the northwest, wars against the Ottomans had an indirect negative influence on the development of the cultural landscape in Istria, foremost during the War of the League of Cambrai (1508–1523) and the Uskok conflicts between the Habsburg Monarchy and the Republic of Venice. During the first years of conflict, Austrian (para)military units, primarily Uskoks, pillaged and destroyed agricultural holdings and rural settlements belonging to the Republic of Venice in the interior of Istria—especially around the city of Buzet (Budak 2007). Thereby, border areas of the Republic of Venice in Istria's interior, primarily those along the border of the Duchy of Pazin, were emptied of population and livestock.

With the establishment of peace, the Venetian authorities again implemented their policy of development of agricultural land and cultivation of various grains (wheat, barley, rice, oats, millet, spelt, and vegetables) in the southern and western parts of Istria. Despite the increase in production, grains still had to be imported. The political and economic centralism that characterised the Republic of Venice smothered every economic initiative of Istrian cities, and systematically prevented all attempts to stimulate the Peninsula's economy, even in situations where local interests did not compete with the Venetian interests. The consequence of this was population regression in Istrian cities (Bertoša 2007).

The first manufacturing facilities in Istria were opened under the Venetian rule. Like their counterparts in Dalmatia, they were based on local raw materials: textiles (Buzet and Poreč) and leather (Rovinj). Poreč and Rovinj were also the main shipbuilding centres. Mined minerals were exploited in the form of sulphur (from the Mirna Valley), rock sand (from the surroundings of Pula), as well as building stone (from the surroundings of Pula, Rovinj, and Poreč). In relation to Dalmatia, however, production output was humble.

War, as well as plague epidemics and the spread of malaria, was negatively reflected in population trends. This is best shown by the drop in population from around 70,000 (1580) to around 35,000 (1620). It took until the end of the eighteenth century for Istria to recover its war-time population losses and exceed 80,000 inhabitants (Gelo 1987).

7.9 The Urban Network

New economic and political circumstances required new territorial organisation, and the development of new urban networks in Croatian territory played an important role therein. Cities in the continental part of Croatia underwent the most dynamic development, especially those located along the Pannonian–north Adriatic transportation system. Trieste, the main harbour for the export of Croatian products, grew to be many times larger than other eastern Adriatic harbour cities during this era.

Rijeka was an important centre in the Pannonian–north Adriatic transportation system and had 5,000–10,000 residents. Other cities, located outside the system, like Varaždin, Zagreb,[20] and Osijek (in continental Croatia), and Zadar, Split, and Dubrovnik (on the Adriatic coast) were of similar size. Senj, Bakar, and Kraljevica (on the northern Adriatic coast), and the river harbours Sisak (Sava and Kupa rivers), Karlovac (Kupa River), Vukovar (Danube River), Koprivnica, and Križevci were secondary centres with roughly 5,000 inhabitants each (Rogić 1990).

It should be noted that a gradual shift in urban development in peri-Pannonian and Pannonian parts of Croatia began during the eighteenth century. This is due to the change in the focus of maritime trade from the Mediterranean Sea to the Atlantic Ocean, whereby Adriatic harbours lost much of their trade volume–the key impetus for economic, population, and urban development. The stumbling development of Dubrovnik compared to Istrian and Dalmatian harbours within the Republic of Venice—which were additionally burdened by the oppressive Venetian economic policy—shows this well.

Although the interior of Dalmatia, in contrast to the interior of Istria, was characterised by smaller urban centres—largely due to the strategic significance and systematic settlement of the Dalmatian Hinterland—they did not experience significant development. Austrian authorities had similar problems in Lika and Gorski Kotar, where, despite best efforts, they weren't able to develop any larger urban settlements (Rogić 1990).

7.10 Settlement and Landscape Change as a Consequence of the Ottoman Territorial Retreat

(Re)settlement of emptied parts of Croatia with Catholic and Orthodox Christians from Bosnia and Herzegovina, after the suppression of the former Ottoman state apparatus, would mark a new turn for the area's cultural landscape. The largest problem that the Habsburg and Venetian governments met stemmed from the very low level of development of agricultural techniques and skills of the newly arrived population. Historical documents testify to this and state how the new settlers only knew how to handle livestock, and were not overly familiar with advanced crop cultivation techniques. The Habsburg authorities immediately addressed this problem during colonisation of the population from the Eyalet of Bosnia into the Pannonian Basin. Their answer was to break up the newcomers into smaller groups and put them into carefully chosen areas. The idea was to put newcomers near pockets of existing inhabitants who could teach them advanced crop cultivation techniques.

Settlers from more (agriculturally) advanced parts of Europe, such as Germans, Austrians, Czechs, Slovaks, and Hungarians, played a significant role due to their experience and skills. Newly arrived settlers from the Ottoman Empire would learn advanced land working techniques from both the existing Croatian population and other newcomers from the north. The best results

[20]In 1750, Zagreb was linked to Karlovac by a modern road and thereby also linked to the Pannonian–north Adriatic transportation system.

in converting the new colonists to a sedentary lifestyle were achieved by the Habsburg authorities in the eastern part of the Pannonian Basin, where there were also the best conditions for the development of crop cultivation. In this area, there was also a significant presence of Central European ethnic elements. It is, therefore, no surprise that this is the area of Croatia where the best outcomes in terms of making a new cultural landscape were achieved. The exceptions to this trend were individual mountainous areas in central Slavonia (Papuk, Psunj, and Ravna Gora mountains) where the newly settled population continued to practice traditional forms of polyculture and herding.

This situation was reversed in hilly and karst areas (Lika, Kordun, and Gorski Kotar) where the land was not as good, nor were there significant groups of settlers from Central Europe. Thus, weaker outcomes were achieved in developing a new cultural landscape, despite significant efforts.

In contrast to the Habsburg leadership, which was absolutist and oriented toward the Central European Germanic cultural circle, different development tendencies prevailed in Venice. Namely, the Venetian authorities allowed the spontaneous settlement of Catholic and Orthodox populations from the Eyalet of Bosnia into their newly acquired holdings (*Acquisto nouvo* and *Acquisto nuovissimo*). No efforts were made to break settlers up into smaller groups, rather entire settlements would often move into Venetian Dalmatia and set up in a new location together. Furthermore, no efforts were made to settle newcomers near existing populations in order to expose them to advanced farming techniques. The Venetian authorities spent some effort searching for farmers (from the Apennine Peninsula) who would be willing to resettle into Dalmatia, but quickly gave up. No one from Italy wanted to live that close to the Ottoman Empire.

The karst environment also required much more effort to be invested into it in terms of land cultivation and agricultural development—especially crop cultivation—than in the Pannonian Basin. However, rather than aiming for specific goals and investing effort accordingly, the Venetian authorities did not show much interest in the general development and advancement of their acquisitions in Dalmatia. This was largely due to the collapse of the Venetian state's corrupt administrative apparatus. This stemmed from the fact that the Venetian political system was, in large part, still based on Medieval feudalism; in contrast to that of the Habsburg Monarchy, which had a system that showed itself to be capable of incorporating new political concepts. Therefore, repatriation of the Vlach population—dissatisfied with the Venetian rule in Dalmatia—back into the Eyalet of Bosnia comes as no surprise; e.g. in 1684, 200 Vlach families returned to the Eyalet of Bosnia (Borković 1982).

The incompetence of local administrators combined with the colonialistic relationship between Venice and Dalmatia—the former regarded the latter only as a place for resource exploitation—came to a head, and the newly arrived settlers were left to fend for themselves. Instead of learning new and advanced crop cultivation techniques and new trades, they spent their time finding ways to dodge their obligations toward their landlords. Distrust of landlords was closely linked to small rural communities becoming closed to the outside, which had the additional effect of preventing the spirit of entrepreneurship and innovation from taking hold (Novak 1944). The underlying reasons behind the area's problems, however, cannot be disregarded; two centuries of war between the Ottomans and Venetians, as well as continuous raiding in Dalmatia by various paramilitary groups (Akinjis, Martolos, Uskoks, and Hajduks), famine, poor governance, and weak natural conditions for agriculture, led to multiple waves of population emigration followed by settlement of new populations. There was little continuous development throughout Venetian Dalmatia as a result.

Venetian Istria, however, was a different story. This was due to the fact that there was no significant settlement during the seventeenth century, as well as the Peninsula's superior natural conditions for crop cultivation. Istria was unique in that it was divided in half politically between the Republic of Venice's coastal holdings and the Habsburg Monarchy's territory in the Peninsula's interior (The Duchy of Pazin). Competition between these two states spurred its two respective governments to implement various measures that were designed to attract and retain settlers from other parts of Croatia. This enabled developmental continuity and the development of an advanced cultural landscape.

References

Bertoša M (2007) Istarsko rano novovjekovlje: razvojne smjernice od 16. do 18. st. In: Budak N (ed) Dalmacija, Dubrovnik i Istra u ranome novom vijeku. Zagreb, Leykam international, pp 80–120

Blanc A (1953) Odnos geografije i historije prikazan na primjeru zapadne Hrvatske. Geografski Glasnik 14–15:35–45

Borković V (1982) Historijsko-geografske osnove naseljavanja Cetine. Geografski glasnik 44:69–84

Budak N (2007) Hrvatska i Slavonija u ranome novom vijeku. Leykam international, Zagreb

Busch-Zantner R (1938) Agrarverfassung, Gessellschaft und Siedlung in Südosteuropa: Unter bes. Berücks. d. Türkenzeit, Harrassowitz, Leipzig

Buzov S (1992) Vlaško pitanje i osmanlijski izvori. Povijesni Prilozi 11:39–60

Buzov S (1993) Razgraničenje između Bosanskog pašaluka i mletačke Dalmacije nakon kandijskog rata. Povijesni Prilozi 12:1–38

Ćosić S (2012) Topografske karte hrvatskih zemalja do početka 20. stoljeća. In: Frangeš S (ed) Topografske karte na području Hrvatske. Zagreb, Državna geodetska uprava, pp 19–55

Ćosić S, Vekarić N (2003) Hrvatski jug: Dubrovačka Republika i Boka kotorska. In: Golub I (ed) Hrvatska i Europa – kultura, znanost i umjetnost: svezak III. Barok i prosvjetiteljstvo (XVII – XVIII. stoljeće). Školska knjiga, Zagreb, pp 79–93

Dugački V, Regan K (eds) (2018) Hrvatski povijesni atlas. Leksikografski zavod Miroslav Krleža, Zagreb

Foretić D (1980) Povijest Dubrovnika do 1808., drugi dio – od 1526. do 1808. Nakladni zavod Matice hrvatske, Zagreb

Fuerst-Bjeliš B (1998) Triplex Confinium – an eco-historic draft. In: Roksandić D (ed) Microhistory of the Triplex Confinium: international project conference papers, Budapest, March 1997. Institute on Southeastern Europe – Central European University, Budapest, p 147

Fuerst-Bjeliš B (2003) Reading the Venetian cadastral record: an evidence for the environment, population and cultural landscape of the 18th century Dalmatia. Hrvatski Geografski Glasnik 65(1):47–62

Fuerst-Bjeliš B, D'Alessio G, Diklić O (2003) Mletački katastar 18. st. i ekohistorijska evaluacija tromeđe. In: Cambi N (ed) Triplex Confinium (1500–1800): ekohistorija: zbornik radova, Split – Zagreb, 2000

Fuerst-Bjeliš B, Kale J (2018) Povijesna dinamika kulturnih krajolika šibenskog područja. In: Kurelac I (ed) Šibenik od prvog spomena: zbornik radova s međunarodnog znanstvenog skupa 950 godina od prvog spomena Šibenika, Šibenik, September 2016. HAZU, Šibenik-Zagreb, p 546

Fuerst-Bjeliš B, Zupanc I (2007a) Images of the Croatian borderlands: selected examples of early modern cartography. Hrvatski Geografski Glasnik 69(1):7–23

Fuerst-Bjeliš B, Zupanc I (2007b) New 18th century Venetian Border in Croatia and its spatial and demographic implications. Hrvatski Geografski Glasnik 69(2):41–52

Gelo J (1987) Demografske promjene u Hrvatskoj od 1780. do 1981. godine. Globus, Zagreb

Glamuzina N, Fuerst-Bjeliš B (2015) Historijska geografija Hrvatske. Sveučilište u Splitu, Split

Gruenfelder AM (2003) Velebitske šume od 16. do 18. stoljeća. In: Cambi N (ed) Triplex Confinium (1500–1800): ekohistorija: zbornik radova, Split – Zagreb, 2000

Holjevac Ž (2003) Ljudi, voda i prirodna sredina na Triplex Confiniumu: Vukasovićeve i Hacquetove ekohistorijske opservacije u Lici i Krbavi u drugoj polovici 18. stoljeća. In: Cambi N (ed) Triplex Confinium (1500–1800): ekohistorija: zbornik radova, Split – Zagreb, 2000

Holjevac Ž (2007) Hrvatsko-slavonska Vojna krajina u ranome novom vijeku. In: Budak N (ed) Hrvatsko-slavonska Vojna krajina i Hrvati pod vlašću Osmanskoga Carstva u ranome novom vijeku. Leykam international, Zagreb, pp 6–107

Inalcik H (2002) Osmansko Carstvo: klasično doba 1300.-1600. Srednja Europa, Zagreb

Iveljić I (2010) Banska Hrvatska i Vojna krajina od prosvjećenog apsolutizma do 1848. godine. Leykam international, Zagreb

Karger A (1963) Die Entwicklung der Siedlungen im westlichen Slawonien: ein Betrag zur Kulturgeographie des Save-Drau-Zwischenstromlandes. Kölner Geographische Arbeiten, heft 45, Franz Steiner Verlag, Wiesbaden

Kaser K (1997) Slobodan seljak i vojnik: I. Rana krajiška društva (1545–1754), II. Povojčeno društvo (1754–1881). Naprijed, Zagreb

Kaser K (2003) Uništenje šuma na obalnom krškom području hrvatske Vojne krajine u prvoj polovici 18. stoljeća: njegovi demografski, privredni i socijalni uzroci. In: Cambi N (ed) Triplex Confinium (1500–1800): ekohistorija: zbornik radova, Split – Zagreb, 2000

Kaser K, Grandits H, Gruber S (2003) Popis Like i Krbave 1712. godine: obitelj, zemljišni posjed i etničnost u jugozapadnoj Hrvatskoj. Srpsko kulturno društvo Prosvjeta, Zagreb

Krivošić S (1990) Stanovništvo Dubrovnika i demografske promjene u prošlosti. Zavod za povijesne znanosti JAZU u Dubrovniku, Dubrovnik

Matuz J (1992) Osmansko Carstvo. Školska knjiga, Zagreb

Mažuran I (1988) Popis naselja i stanovništva u Slavoniji 1698. godine. Zavod za znanstveni rad u Osijeku JAZU, Osijek

Merlić E (2008) Buzeština kroz povijest. Vlastita naklada, Pazin – Buzet

Mirdita Z (2009) Vlasi – starobalkanski narod (od povijesne pojave do danas). Hrvatski institut za povijest, Zagreb

Moačanin N (1998) Introductory essay on an understanding of the Triplex-frontier area: preliminary turkologic research. In: Roksandić D (ed) Microhistory of the Triplex Confinium: international project conference papers, Budapest, March 1997. Institute on Southeastern Europe – Central European University, Budapest, p 125

Novak G (1944) Prošlost Dalmacije, knjiga druga: od Kandijskoga rata do Rapalskoga ugovora. Hrvatski izdavalački bibliografski zavod, Zagreb

Novak G (1961) Povijest Splita: knjiga druga – od 1420. do 1797. god. Matica hrvatska, Split

Pavić R (2012) Geografija Gorske Hrvatske: prilozi analizi socijalno-geografske strukture Gorskoga kotara i Ogulinsko-plaščanske submontane udoline. Vlastita naklada Zdenko Mance, Rijeka

Pederin I (2011) Juraj Matutinović i drugi Napoleonovi savjetnici za Ilirske pokrajine i Bosnu te problemi prodora na jugoistok. In: Trogrlić M, Vrandečić J (eds) Dalmacija za francuske uprave (1806.-1813.): zbronik radova, Split, September 2006. Književni krug – Odsjek za povijest Filozofskog fakulteta u Splitu, Split, p 221

Pejnović D (1985) Srednja Lika – socijalnogeografska transformacija, Centar za kulturu – Muzej Like, Gospić

Pejnović D (2009) Geografska osnova identiteta Like. In: Holjevac Ž (ed) Identitet Like: korijeni i razvitak. Institut društvenih znanosti Ivo Pilar – Područni centar Gospić, pp 45–84

Peričić Š (1998) Gospodarska povijest Dalmacije od 18. do 20. stoljeća. Matica hrvatska Zadar, Zadar

Regan K (ed) (2003) Hrvatski povijesni atlas. Leksikografski zavod Miroslav Krleža, Zagreb

Rogić V (1953) Senj: prilog poznavanju položaja i regionalne funkcije. Geografski Glasnik 14(15):47–64

Rogić V (1954) Senj - historijsko značenje i suvremeno mrtvilo jedne naše luke. Pomorstvo 9(3):134–136

Rogić V (1956) Razlike Pejzaža Velebitskih Padina. Geografski Glasnik 18:15–32

Rogić V (1957) Geografski osnovi stočarskih veza Like i Dalmacije. In: Lisac A-Lj (ed) Zbornik naučnih i književno-umjetničkih priloga bivših đaka i profesora Zagrebačke klasične gimnazije o 350-godišnjem jubileju 1607.-1957., Zagreb, September 1957. Odbor za proslavu 350-godišnjice Klasične gimnazije u Zagrebu, Zagreb, pp 705–714

Rogić V (1958) Velebitska primorska padina – prilog poznavanju evolucije krškog pejzaža (nastavak). Geografski Glasnik 20:53–110

Rogić V (1969a) Rapska Otočna Skupina. Geografski Glasnik 31:109–124

Rogić V (1969b) Les differences qualitatives de la regionalisation a grande et petite echelle – L exemple de la Croatie. Bull De La Faculte Des Lettres De Strasbourg 47:431–436

Rogić V (1970) Prostor riječke komune: prilog poznavanju regionalno-geografske problematike. Geografski Glasnik 32:137–159

Rogić V (1971) Cresko-lošinjski prostor. Zbornik Za Narodni Život I Običaje 45:635–649

Rogić V (1973) Saobraćajna povezanost Like i tranzitne veze Like s ostalim dijelovima zemlje. In: Gušić B (ed) Lika u prošlosti i sadašnjosti: zbornik 5. Historijski arhiv u Karlovcu, Karlovac, pp 631–641

Rogić V (1975a) Osnove historijsko-geografskog razvoja. In: Cvitanović A (ed) Geografija SR Hrvatske – knjiga V: Sjeverno hrvatsko primorje. Školska knjiga, Zagreb, pp 42–45

Rogić V (1975b) Regionalno-geografska izdvojenost i geografske osobine riječkog prostora. In: Cvitanović A (ed) Geografija SR Hrvatske – knjiga V: Sjeverno hrvatsko primorje. Školska knjiga, Zagreb, pp 79–97

Rogić V (1977) Regionalizacija Gorske Hrvatske. Geografski glasnik 39:25–43

Rogić V (1981) Raumgebundene Konstanten der Entwicklungsachsen Donau Adria, Muenchener Studien zur Sozial und Wirtschaftsgeographie, vol 21. Verlag Michael Lassleben Kallmünz, Regensburg

Rogić V (1982) Auswirkungen der oesterreichischen Militärgrenze auf die dinarisch-pannonischen kulturlandschaft. Schriftenreihe Des Regensburger Osteuropa Instituts 8:41–55

Rogić V (1990) Regionalna geografija Jugoslavije, knjiga 1 – prirodna osnova i historijska geografija. Školska knjiga, Zagreb

Rogić V (1991a) Hrvatska vojna krajina – temeljna historijsko-geografska problematika. In: Crkvenčić I (ed) Političko-geografska i demografska pitanja Hrvatske. Savez geografskih društava Hrvatske, Zagreb, pp 165–186

Rogić V (1991b) Croatian Military Border-fundamental historical-geographical problems. Geogr Papers 8:167–189

Rogić V (1992) Problematika regionalne organizacije jugoistočnog segmenta Središnje Hrvatske. Acta Geogr Croatica 27:95–106

Slukan Altić M (2003) Komparativna analiza kulturnog pejsaža ruralnih naselja mletačke i habsburške krajine. In: Cambi N (ed) Triplex Confinium (1500–1800): ekohistorija: zbornik radova, Split – Zagreb, 2000

Slukan Altić M (2006) Povijesna geografija Kosova. Golden marketing – Tehnička knjiga, Zagreb

Soldo JA (2011) Sinjska krajina u 17. i 18. stoljeću. Ogranak Matice hrvatske u Sinju, Sinj

Štefanec N (2003) Trgovina drvetom na Triplex Confinium ili kako izvući novac iz senjskih šuma (1600–

1630)? In: Cambi N (ed) Triplex Confinium (1500–1800): ekohistorija: zbornik radova, Split - Zagreb, 2000

Valentić M (2003) Geografska i demografska slika Hrvatske u XVII. I XVIII. stoljeću. In: Golub I (ed) Hrvatska i Europa - kultura, znanost i umjetnost: svezak III. Barok i prosvjetiteljstvo (XVII - XVIII. stoljeće). Školska knjiga, Zagreb, p 29–42

Vrandečić J (2007) Regio maritima: Dalmacija i Dubrovnik u ranome novom vijeku (1420. - 1797.). In: Budak N (ed) Dalmacija, Dubrovnik i Istra u ranome novom vijeku. Zagreb, Leykam international, pp 7–79

Yriarte C (1883) La rive dell' Adriatico e il Montenegro. Venezia, Chioggia, Trieste, l'Istria, il Quarnero e sue isole, la Dalmazia, il Montenegro, Ravenna, Ancona, Loreto, Foggia, Brindisi, Lecce, Otranto. Fratelli Treves, Milano

8 Spatial Impacts of Industrial Development Within the Austro-Hungarian Monarchy

Abstract

The most important change that affected Croatia at the turn of the eighteen century to the nineteenth century was the dissolution of the Republic of Venice (1797), and the subsequent division of Venetian territory between France and Austria. This resulted in the Habsburg Monarchy taking control of Istria and the parts of Dalmatia formerly controlled by Venice, putting the entirety of contemporary Croatia under its rule. The short period of French administration in the western parts of Croatian territory was marked by important development projects. After the Austro-Hungarian Compromise of 1867, Croatian territory was divided between Austria and Hungary, leading to an increasing political struggle for (and against) the unification of Croatia. The politics of railway construction throughout Croatian territory, which was largely determined by Vienna and Budapest, would dictate the transport and economic orientation of Croatia along the Central European–north Adriatic axis. Industrialisation stimulated the dynamic economic and population development of cities in these areas of Croatia. Zagreb became the main centre of urban development and turned into a modern national economic centre with important cultural, educational, and political functions. In contrast, the end of the nineteenth century was a time of great overseas migration waves, mostly from rural Dalmatia, as a consequence of the grape vine diseases at the peak of the wine conjucture.

Keywords

Croatia · Austro-Hungarian Monarchy · Railroad network · Industrialisation · Urbanisation · Migration waves · Zagreb

Turbulent political events at the turn of the eighteenth to the nineteenth century followed the French Revolution (1789), shaking European politics to the core. They also left their mark on Croatia, with the most important event (for Croatia) being the dissolution of the Republic of Venice in 1797 and the subsequent division of Venetian territory between France and Austria. This resulted in the Habsburg Monarchy taking control of Istria and the parts of Dalmatia formerly controlled by Venice, putting the entirety

B. Fuerst-Bjeliš and N. Glamuzina, *The Historical Geography of Croatia*,
Historical Geography and Geosciences, https://doi.org/10.1007/978-3-030-68433-4_8

The Croatian State Archives, Zagreb
The Croatian State Archives is located in the most significant Croatian *art nouveau* palace. Built from 1911 to 1913 as the Royal University Library and the Royal State Archives, the palace with its gardens was a part of the modern town planning model of orthogonal alteration of the squares and parks (Photograph: Ivana Erdelez. Reproduced with permission)

of contemporary Croatia under its rule. The newly acquired areas were placed under special administration by Habsburg authorities, which prevented the political unification of all Croatian territories within Habsburg territory. During the short-lived first iteration of Austrian rule over Dalmatia and Istria, no significant changes took place in the cultural landscape.

8.1 The Political–Geographical Context of Change Under Austria–Hungary

The Monarchy was forced to abandon its newly gained territories (from Venice) in 1805 to France, which would fortify the area in the years to come. France also used the opportunity to take The Republic of Ragusa (Dubrovnik) in order to prevent Russo-Montenegrin forces from striking north from the occupied Bay of Kotor. After French military authorities dissolved the Republic of Ragusa (Dubrovnik) in 1808, its former territory was reannexed into Dalmatia, after nearly three centuries of separation.

With the new peace treaty in 1809, the Habsburg Monarchy surrendered the southern part of Banovina and the Karlovac Military Frontier to France. These territories, along with former Venetian Dalmatia and Istria (as well as Carniola, Gorizia, Carinthia, and part of Tyrolia), would become part of the Illyrian Provinces (*Provinces illyriennes*) that were overseen from Ljubljana (Fig. 8.1).

French authorities, in contrast to their Habsburg counterparts, attempted to undertake wide-reaching political, social, and economic reforms, in order to advance the new territories under their rule (Croatia included) in keeping with the ideals of the Revolution. With this goal in mind, they made a major effort to develop Croatian territory under their rule, and their efforts were most visible in Dalmatia and Lika. The results were, however, rather humble in the great scheme of things, as it wasn't possible to sow the seeds of the Revolution in a society where the bourgeois were very weakly developed. Due to this, apart from paved roads Zadar–Dubrovnik, Via Louisiana from Karlovac to Rijeka, and the road from Pula to Trieste via Pazin—there were no significant changes to the cultural landscape.

French authorities incentivised grape and olive production, fruit orchards, potatoes, herding, fishing, and ended limitations on tobacco production. New manufacturing facilities that used local raw materials to produce bricks (Zemunik near Zadar, Dobrinj on the island of Krk, Opuzen in the Neretva Delta), wax (Split), and lime (Drniš in the Dalmatian Hinterland) were also opened. Furthermore, rational forestry policies were also enacted, which was important in Dalmatia—where the forests were especially threatened.

The reasons that the French economic policies did not succeed in taking root in Croatian territory can largely be found in the uncertain political conditions in Europe and the lack of available capital. An additional problem was the fact that the British were blockading the European continent during this era (1806–1814), which prevented maritime trade with other parts of the world. This hit Croatian territory under French rule heavily as the coastal areas depended heavily on maritime trade. Inherited problems regarding economic (under)development spurred migrations between individual parts of Croatia. Migrations from Dalmatia to Slavonia and even Bosnia[1] were particularly frequent.

The project of the Illyrian Provinces flopped because there were almost no economic links between its constituent areas. The lack of cohesion meant that these areas largely functioned independently of one another. Significant distances between individual areas and weakly developed transport links severely hindered economic, political, and social integration. Although French authorities attempted to solve this problem by building roads and promoting economic development, the exceptionally weak interactions (in terms of flows of people, goods,

[1]Dalmatia continued to be struck by famine, the worst of which took place in 1807 and 1811–1812.

Fig. 8.1 Croatian territory within the Illyrian Provinces under Napoleonic rule (1805–1813) (*Source* Regan 2003)

and information) among the Provinces' constituent areas did not significantly improve. Short-term French rule, during which the entire Croatian Littoral, along with Lika, Gorski Kotar, and part of the peri-Pannonian belt south of the Kupa and Sava rivers were united into a single administrative unit, ended in 1814, when the area of the Illyrian Provinces again came under Habsburg rule.

8.2 Political and Regional Divides Within the Austro-Hungarian Monarchy

The area of contemporary Croatia was, during the era of Austro-Hungarian rule, divided into many administrative/political units (Fig. 8.2). Each of them had their own provincial

Fig. 8.2 The Austro-Hungarian Monarchy within Europe during the second half of the nineteenth century (*Source* Regan 2003)

government that answered to higher state authorities in either Vienna or Budapest. Economic development of each political unit took place during the process of the shaping of the modern nation-state and was often hindered by problems with nationalism and the hegemonic policies of the Austrian and Hungarian governing bodies.

When the Austro-Hungarian Empire came into being in 1867, Croatian territory was divided into six individual administrative units, based on the Austro-Hungarian Compromise of 1867 and the Hungarian-Croatian Compromise of 1868 (Fig. 8.3).

Thereafter, the following parts of Croatia belonged to the Hungarian part of the Monarchy and its seat in Budapest:

1. The Kingdom of Croatia and Slavonia, consisting of nearly the entire Croatian part of the Pannonian Basin (apart from the Military Frontier), Mountainous Croatia, and the Kvarner Littoral with the exception of Rijeka.
2. The southern part of Baranja (within contemporary Croatia) and Međimurje were part of the Kingdom of Hungary along with Erdelj.
3. The harbour of Rijeka was separated from the rest of Croatian territory via *corpus separatum* and was under the direct jurisdiction of authorities in Budapest (Fig. 8.4).

Within the Austrian part of the Dual Monarchy, under the direct rule of Vienna, were predominantly littoral areas, but also individual parts of continental Croatia that made up the Military Frontier:

1. *Küstenland* was a region in Croatia which was also referred to as the "Austrian Littoral". It was made up of the Croatian part of Istria and the islands Krk, Cres, and Lošinj.
2. Dalmatia, which consisted of the entire contemporary region of the same name and the island of Rab (which is not a part of contemporary Dalmatia).

Fig. 8.3 Administrative-territorial division of Austria–Hungary from the second half of the nineteenth century to the end of World War I (*Source* Population Census in Austria-Hungary (1868–1910), Chech Statistical Office, www.czso.cz, accessed Nov. 26th, 2019)

3. The Military Frontier, which consisted of a large part of continental Croatia: the Lika Region; most of Gorski Kotar; the Kvarner Littoral; Kordun; Banovina; the southern part of the area along the Kupa and Sava rivers; Moslavina; parts of western Slavonia and Podravina, and the far eastern part of Srijem. The Military Frontier was administered directly by Viennese authorities. It was dissolved in 1881, and its constituent territories were annexed into the Kingdom of Croatia and Slavonia.

The political division of Croatian territory reflected negatively on economic development, which is evident in the economic stagnation in Croatia during the Industrial Revolution. This had long-term consequences, which left direct and indirect marks on the area.

8.3 Railway Networks—Implications of the Three Railway Concepts for the Territory of Croatia

The era of political reformation of the Austrian Empire into the Austro-Hungarian Empire (1867) coincided with the era of building and widening of railway networks. This spelled the end of the dominance of the Pannonian–north Adriatic transportation system. Railway construction also coincided with the era of the humble "Industrial Revolution". Although the first railways were

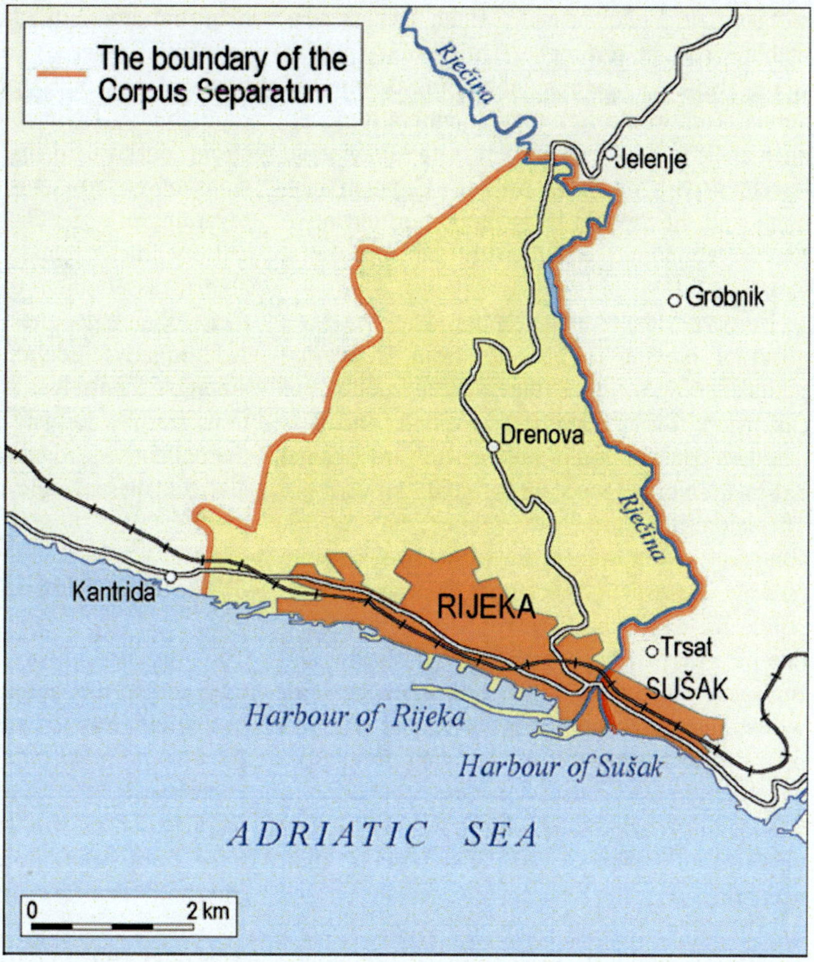

Fig. 8.4 *Corpus separatum*. The harbour of Rijeka under Hungarian rule during the Austro-Hungarian era (*Source* Proleksis Encyclopaedia, The Miroslav Krleža Institute of Lexicography, http://proleksis.lzmk.hr. Accessed 26 Nov 2019)

privately owned, in 1845, authorities of the Empire decided to take over their upkeep and the network became a singularly important element of the Empire's internal political battle (between Vienna and Budapest) and a tool for directing economic development (Rogić 1990).

Construction of the first railways began before the Austro-Hungarian Compromise of 1867. Among the first to be finished (1857) was the southern line (*Südbahn*), which linked Vienna to Trieste—the main harbour of the Habsburg state. This magistral line ran through Graz, Marburg (Maribor), Celje, and Ljubljana, and was built through historically Austrian regions on difficult Alpine slopes. This was the first and best railway linking the interior of Europe with the Mediterranean. The Austrian system of railway construction was based on radial branching of magistral lines outward from Vienna—the main political centre. The radial railway network was meant to be a means to better integrate the Empire, economically and politically. Local lines were linked to the magistral lines, which functioned as the main transit ways. The construction of the first lines toward Croatian territory, toward river harbours on the Sava and Kupa rivers, began in the 1860s. As a direct result, a railway from Zidani Most (in Slovenia) to Zagreb was opened, and was further extended to Sisak and Karlovac. This began the process of redirecting transit from the Pannonian lowlands toward the main Austrian harbour Trieste.

Croatian economic and political circles put forward the idea of constructing railways that would link the Croatian northern Adriatic harbours of Rijeka and Senj with the Danube River harbours of Vukovar and Zemun (which was within Croatian territory at this point), which would also be of use to neighbouring Bosnia and Serbia. This would link the agriculturally developed areas in the southern part of the Pannonian lowlands with northern Adriatic Croatian harbours and thus would be an important incentive for the economic development of continental Croatia.

However, the Austrian plan, which was a result of bowing to pressure from Hungarian political circles, was the official policy of expansion toward Southeast Europe, and was based on construction of railways toward Bosnia, Kosovo, North Macedonia, and onward to Thessaloniki.[2] Furthermore, after the Austro-Hungarian Compromise in 1867, two separate railway construction schemes emerged, both politically motivated. The result was two separate railway networks, with hubs in Vienna and Budapest, respectively—the Hungarians network was similar in concept to its Viennese counterpart, but with Budapest as the main point from which rail lines radiated. After the occupation of Bosnia and Herzegovina (1878), individual parts of Croatian territory were linked via a third network.

All three railway networks were mainly results of political manoeuvering—economic development was a secondary concern. Economic motivations for railway construction were closely linked and inseparable from the official policies of Vienna and Budapest. This meant that the three separate railway systems with two main hubs (Vienna and Budapest) took shape (Rogić 1990). This would have a significant impact on Croatian territory.

The failure of Croatian political and economic circles to realise their plan of constructing railways that would be ideal for linking and economically developing Croatian territory, stems from two main problems. First, Croatian political and economic circles were politically weak, and second, they often worked at cross-purposes. It is a fact that the construction of railways, especially magistral lines, did not exclusively depend on economic criteria, i.e. on the availability of financial resources for construction and future profitability. Their construction was ordered by strong political forces in keeping with the decisions of governments in Vienna and Budapest regarding the general acceptability of a given railway line. Croatian political circles were not able to assert their will against the strength of their Viennese and Budapester counterparts. Additionally, individual Croatian cities fought amongst themselves over the selection of locations for new railway lines, further undermining the collective Croatian position vis-à-vis Vienna and Budapest (Rogić 1990; Iveljić 2010).

8.3.1 The Austrian Railway Network

The Austrian railway network in Croatian territory under the administration of Vienna was primarily a way to help build the navy and affirm Pula as the main military harbour. Due to these concerns, Pula was linked with Trieste via the main Istrian railway line (*Istrianerbahn* or *Istrianer Staatsbahn*) in 1876, while the line from Pula to Buzet (also finished in 1876) was only significant on the local level. Poreč was linked to Trieste in 1902 via a special narrow-tracked line that went through Savudrija, Buje, and Motovun.

In contrast, the construction of the railway network in Dalmatia, called the Dalmatian line (*Dalmatinerbahn*), was limited from its inception to linking Split and Šibenik with the important coal mines in Siverić (1877). This line would later be extended to Drniš and Knin (1888) in the Dalmatian Hinterland, while Split would be linked with Sinj by a narrow-tracked line (1903). The latter line was supposed to be the first part of a narrow-tracked line that would link the Dalmatian, Bosnian, and Pannonian areas (Split–Sinj–Bugojno–Sarajevo–Brod on the Sava (contemporary Slavonski Brod)). This plan, however, never came to pass. This was partly due to political jockeying between Vienna and Budapest and partly because of the uncertain future of Bosnia and Herzegovina. Because this line was never realised, the import of goods from Bosnia

[2] Construction of the railway to Thessaloniki was finished in 1874.

and Herzegovina continued via caravans and along the existing road from Sinj to Split (Novak 1965). The situation was similar for goods coming from western Bosnia and Bihać to the northern Adriatic harbour Senj.

In southern Dalmatia, Metković, an important centre for maritime transport on the Neretva River, was linked to the narrow-tracked Bosno-Herzegovinian network by a narrow-tracked line. Dubrovnik was linked to the same network in 1901, on the line that linked Metković with the military harbour Zelenika in the Bay of Kotor. Zadar, however, was never linked to the railway network while under Austrian rule, despite its role as the political centre of Habsburg-ruled Dalmatia.[3]

In 1907, an agreement between Austrian and Hungarian authorities was reached regarding linking the Slovenian city Metlika with Karlovac, and onward to Knin via Ogulin in Gorski Kotar. This was supposed to connect Dalmatia with the rest of the Empire.[4] Work began in 1913, but was greatly hindered due to the onset of the First World War. Work was stopped in 1918 with the end of the Austro-Hungarian Empire. By the end of construction only a short segment in Gorski Kotar, from Ogulin to Plaški, had been constructed, while the segment from Plaški to Vrhovine was limited in the amount of traffic it could handle (Bunijevac 2009).

From this, we can conclude that the Austrian railway network was built with narrow military goals in mind, and best served the area of Istria. In contrast, Dalmatia remained a sort of *cul-de-sac* under Austrian rule, which the government in Vienna was not overly interested in connecting to the rest of the Empire. This can be easily seen in the fact that Dalmatia, at the start of the twentieth century (1904), was the Austro-Hungarian province with the shortest railway network: only 126 km. Although the network had grown to 230 km (narrow-tracked lines) by the time the Empire fell, it would not be enough to spur meaningful economic development, especially in conditions of transport isolation and underuse of Dalmatian harbours in international trade.

The construction of the Austrian railway system resulted in the most important, for Croatia, railway line along the Sava lowlands, which would link the Pannonian lowlands with the northern Adriatic (Zemun–Rijeka). This line was insisted upon by Croatian economic and political circles and had the simultaneous effect of slowing economic development of the Military Frontier (Rogić 1982).

Additionally, there were plans to remedy the fact that Dalmatia was not linked directly via railway with other parts of the Empire. In order to develop other harbour cities, Dalmatian economic circles intended to build railway links with Bosnia and Herzegovina, and eventually with Serbia—a wide hinterland that could export its products via Dalmatian harbours. The emphasis of local government, especially that of Split, was also on building railway connections, primarily with Zagreb as the main national centre. However, none of these attempts came to fruition due to the political machinations of more powerful circles within the Empire. This meant that all attempts on the part of Split to assert itself as an important Adriatic harbour ultimately failed (Novak 1965).

The advent of steamships helped alleviate (somewhat) the transport isolation of Dalmatian harbours (Split, Šibenik, Metković, Dubrovnik), in relation to their northern Adriatic counterparts (Rijeka, Pula, Rovinj, and Poreč) that were well connected to the railway system. This development bore meagre fruit, however, apart from in Rijeka, which developed into the most important harbour in Croatia in the early era of manufacturing. Efforts by Dalmatian economic circles to obtain the status of "free harbour" for Split failed because the Austrians wanted to protect their main mercantile harbour Trieste from potential competition. Vienna had the support of Budapest in this matter as the Hungarians also wanted to protect the prominent position of their main

[3]Although Zadar gained the status of "open city" in 1870, it never became competition for Trieste or Rijeka because it wasn't connected to any of the three railway networks.

[4]Although Hungarian authorities agreed with the idea of linking Dalmatia with the rest of Austria–Hungary in principle, in practice they hindered the construction of the railway through Lika in order to stifle potential competition for Rijeka—the main harbour under their control.

harbour Rijeka. This sort of economic protectionism also had an ulterior political motivation —to undermine the process of national integration of Croatian territories.

Dalmatia's transport isolation led it to fall behind economically and developmentally. The governmental bodies and economic circles in Vienna (and Budapest) were not interested in linking Dalmatia with the rest of the Empire via railway. Despite this, Viennese leadership had a plan for the construction of the so-called Split line (*Spalatobahn*) in 1894, linking Sarajevo to Bugojno, in order to gain a stronger economic position in Bosnia and Herzegovina. This line would allow them to exploit raw materials which were needed for Austrian industry and enable easier transport of Austrian industrial products to the markets of Bosnia and Herzegovina. The extension of the tracks from Bugojno to Split, however, never happened (apart from a short stretch from Split to Sinj in 1903) due to emergent conflicts of interest between Vienna and Budapest. The Hungarian government regarded linking Western Bosnia with Split as against its interests, seeing the development as a potential threat to the primacy of their harbour Rijeka.

8.3.2 The Hungarian Railway Network

The Hungarian railway network was based on the construction of six magistral lines (Rogić 1990) which extended away from Budapest radially. Two of these six lines led toward Croatia, whereby the link between the Hungarian part of the Danubian Lowlands and the northern Adriatic was of key importance. The first, built in 1873, was a western magistral line that linked Budapest with Rijeka through Koprivnica and Zagreb. This meant the most desirable place to put the railway line, which would have run to Senj and Bakar through Vinodol, was left unused. The reasons for this choice were political and promoted the economic interests of factions in Hungary (e.g. those exploiting the dense forests of Gorski Kotar) as well as the direct Budapest–Rijeka link. This led the Hungarian government to buy out Austrian railway lines which had been built as part of the greater Südbahn line in Croatia. The persistent pressure of their centralisation policy when it came to decisions regarding railway construction was one of the main facets of Hungarian economic policy.

In the east, due to the significance of Croatian territory, there was the central Budapest line through western Voivodina: Budapest–Subotica–Sombor. This line was extended to Dalj (a harbour on the Danube), and on to Vukovar and Vinkovci, as well as Brod on the Sava (Slavonski Brod). Its construction started after the Austro-Hungarian occupation of Bosnia and Herzegovina, and it would later become the main transport link between Hungary (mainly Budapest) and Bosnia and Herzegovina.

This radial railway network's main purpose was to further Hungarian irredentist politics (Greater Hungary). This is shown by the fact that the eastern parts of Slavonia, Srijem, Baranja were linked with Zagreb by railway lines that went through Budapest first, which was rather out of the way (Rogić 1990). As such, Croatia had the function of a transitory area, by which Hungarian political circles attempted to spread their political influence (into Bosnia and Herzegovina), as well as economic goals: the export of raw materials[5] and import of goods from other parts of the world (Juzbašić 1974, 2002).

Hungarian authorities succeeded in redirecting the greater part of Bosnian exports toward Rijeka rather than Trieste at the turn of the nineteenth century to the twentieth century (Juzbašić 1974, 2002). This additionally spurred the development of Rijeka, as well as Brod on the Sava (Slavonski Brod), which had gained an important nodal role in the system of industrial facilities related to maintaining the railways.

[5]Raw materials made up 83% of total Bosno-Herzegovinian exports from 1900 to 1910. Although the amount of material exported was three times larger than the amount of imported material, the monetary value of imports was greater.

8.3.3 The Bosno-Herzegovinian Narrow-Gauge Railway Network

The Bosno-Herzegovinian railway network was built during the era of Austro-Hungarian rule over Bosnia and Herzegovina and was coloured by the dualistic relations within the Empire, as well as by Bosnia and Herzegovina's special status. Therefore, its construction was largely a result of the conflicting interests of the two main poles of political and economic power within the Empire: Vienna and Budapest. This directly affected the Croatian railway network, being that it was connected to the rest of the Austro-Hungarian railway network through Croatian territory. Additionally, it provided Bosnia and Herzegovina a railway connection to the Adriatic Sea in Austrian-ruled Dalmatia.

The construction of a normal-gauge railway had been started while Bosnia and Herzegovina were still under Ottoman rule, meant to connect Sisak with Banja Luka and, thus, with the *Südbahn*. This line was supposed to be the first part in a longer line to Sarajevo and on to Kosovska Mitrovica. Ottoman and Austro-Hungarian leadership worked together to build a normal-gauge line from Banja Luka to the border between Croatia and Bosnia and Herzegovina on the Una River (Dobrljin) in 1872. However, due to the complex relationship between Vienna and Budapest and the indecisiveness of Ottoman authorities when it came to foreign policy, this line was never connected to the west. Moreover, after three years of use, traffic was stopped from 1875 to 1879 due to a rebellion in Bosnia and Herzegovina and general unprofitability.

In contrast, Hungarian circles pushed for two transit lines. The first would not even touch Croatian territory, rather it would start in Budapest and go through Voivodina, Belgrade, and on to the Morava Valley to the south. The second line would start in Slavonia and go to the Danube harbour Dalj, then to Brod on the Sava, connecting with the existing line to Sarajevo. Due to this, a line from Dalj (on the Danube River) through Vinkovci and Vukovar to Brod on the Sava was finished in 1878, as well as the western Croatian part of the Sisak–Dobrljin line (1882), at the border with Bosnia and Herzegovina, completing the link to Banja Luka (Juzbašić 1974, 2002).

Further construction of railway in Bosnia and Herzegovina was undertaken primarily for military-strategic reasons. Due to the need to secure supply lines for Austro-Hungarian military units, which were already active in Bosnia and Herzegovina. Difficulties associated with weakly developed roads, as well as active armed resistance, spurred the decision to build a narrow-gauge railway network. Continuing to build normal-gauge railway tracks was too time and material intensive, so the narrow-gauge Bosnian railway network (*Bosnabahn*) was born.

A narrow-gauge line was easier, faster, and cheaper to build in areas dominated by hilly terrain.[6] The construction of such a network encouraged only a low level of development of Bosnia and Herzegovina, compared to other parts of the Empire. This was due to a lack of financial resources, lack of interest on the part of private entrepreneurs in taking part in building a railway system, and the Austro-Hungarian policy that the only railway that would be built was one for which Bosnia could pay back the cost.

Austro-Hungarian authorities invested funds only in railways that were necessary for military purposes. This resulted in the construction of the first leg of the Bosno-Herzegovinian narrow-gauge line from Brod on the Sava[7] to Zenica (1879) and on to Sarajevo (1882). The completion of these legs was the main incentive for the construction of normal-gauge tracks in Croatian territory through the Sava River Valley from Sunja to Brod on the Sava (Juzbašić 1974, 2002). In 1889, this railway going through the Sava Corridor in Croatian territory was finished, linking Zagreb and Brod on the Sava.

[6]They were informed by past experience with neighbouring Serbia, where a narrow-gauge railway network had been built for the same reason.

[7]Brod on the Sava is actually comprised of two cities straddling the Sava River: contemporary Slavonski Brod and Bosanski Brod. Slavonski Brod is on the Croatian side of the river (left bank), and Bosanski Brod is on the Bosnian side (right bank).

Austro-Hungarian administration, under the leadership of Benjamin Kállay (1882–1903), attempted to strengthen the economic development of Bosnia and Herzegovina by building a narrow-gauge line from Sarajevo to Metković (the so-called Neretva line—*Neretvabahn* or *Narentabahn*). This link gave Metković an important nodal role in railway and maritime transport.

The construction of the line known as the *Zelenikabahn*, from the Herzegovinian settlement Gabela on the Neretva River to Zelenika in the Bay of Kotor (1901)—with a link to Dubrovnik—helped increase the transport of goods through Dalmatian harbours. Metković and Dubrovnik, however, only took a small amount of the total freight transport from neighbouring Bosnia and Herzegovina. This was still meaningful for Metković, which had previously only been a transport point for Bosnian goods that were too far from normal transport lines, and found their way to the Adriatic in the south. The inability of Metković's harbour to accept larger ships limited its potential development. Dubrovnik took over even less transport than Metković, and strong development of Dubrovnik's harbour—as was the case in Metković—was prevented by Hungarian political machinations aimed to preserve Rijeka's superiority. Despite this, the gains made by Metković and Dubrovnik as a result of becoming connected with their respective hinterlands represented a significant boon locally in terms of development.

The other portion of Bosno-Herzegovinian import and export, which took place at Dalmatian harbours that were not linked to the railway network (Split, Šibenik, Zadar), was redirected toward northern Adriatic harbours, primarily Rijeka. The reduction of transport of Bosnian goods through Dalmatian harbours dates from 1879, when the line from Brod on the Sava to Zenica in Bosnia was finished, and again in 1882 when the line was extended to Sarajevo (Juzbašić 1974, 2002).

One small portion of Bosno-Herzegovinian exports continued to go through Šibenik's harbour, thanks to a narrow-gauge line running from Knin (in the Dalmatian Hinterland) to Drvar in western Bosnia. This line created a link to Dalmatia for the western Bosnian narrow-gauge railway network (the so-called *Steinbeisbahn*[8]). This line, however, was primarily used for exporting timber from Western Bosnia; and Otto von Steinbeis financed and adapted the harbour in Šibenik for this specific purpose.

Worsening political conditions in Southeast Europe at the beginning of the twentieth century resulted in the resurrection of the plan to link Bosnia and Herzegovina with Serbia and Montenegro, out of military and strategic concerns. This meant that the planned construction of a narrow-gauge line linking Bosnia and Herzegovina with the harbour in Split would never come to fruition. Starting in 1906, piecemeal construction of railway lines heading southeast took place: Sarajevo–Uvac (in Sandžak) and Višegrad–Vardište (on the border of Bosnia and Herzegovina and Serbia). The line to Vardište would gain importance during the interwar period with a new link to Serbia (Juzbašić 1974, 2002).

8.4 Industrialisation, Urbanisation, and Regional Differentiation of Economic Development

The railway era in Croatia overlapped with changes in the socio-political system and economic conditions on the European and world markets. These directly and indirectly influenced the cultural landscape, as well as many societal processes—foremost population growth and urbanisation. However, to understand the changes in the cultural landscape, three important events that dictated the development of Croatian territory during the second half of the nineteenth century should be mentioned.

First, the formal end of feudal relations on April 25, 1848 was declared by Ban Josip Jelačić. Even though serfdom had been formally

[8]It was named for Otto von Steinbeis (1839–1920), a German businessman and investor who gained a concession to exploit the forests of western Bosnia in 1893 and financed the construction of the railway for this purpose.

ended in 1781, via order of Emperor Josef II, isolated remnants of feudal relations persisted. 1781 is understood as the year that marked the beginning of the end of feudalism, but many elements thereof persisted until the end of the nineteenth century—in the Croatian Littoral, some persisted until the end of Austria–Hungary itself. The change from one social system to another was a long-lasting process that, in Croatia (like in many other parts of the world), provided the impetus for many other social processes.

The second important process was the adoption of capitalist social relations (the exchange of money for goods/services) in all economic areas. The first hints of capitalism emerged and spread quickly, primarily in large urban areas. These were the humble beginnings of the Industrial Revolution, which led to the construction of factories and industrial districts: new elements in urban areas. Capitalist social relations would have the greatest impact on the cultural landscape in rural areas, because they had to change the most to adjust to a market-oriented type of economy.

The third is the dissolution of the Military Frontier in 1881 and the inclusion of its constituent lands into Croatia Proper (*Banska Hrvatska*). Although this didn't have a direct impact on the cultural landscape, it enabled capitalism to penetrate further and faster into an area that was characterised by weaker economic development in relation to other parts of Croatia. This is shown by the fact that 99.28% of the population of the Military Frontier was active in primary sector activities (Rogić 1990; Iveljić 2010). Namely, along with farming, herding and exploitation of forests (for timber) were the dominant activities. The integration of the Military Frontier into Croatia Proper was a fundamental event that spurred the subsequent rapid social and economic transformation, and left its mark on the cultural landscape.

At the start of the nineteenth century, Croatia was a very backward land in social and economic terms—it hadn't changed that much since the Middle Ages. Agriculture, as the main economic activity, was still practiced using old tools (mostly made of wood, with the odd iron plow), and with the aid of animals and a considerable amount of manpower. Under such conditions, dominated by autarchic production, there was practically no market surplus.

Weak development of manufacturing, which was gradually strengthening throughout Western Europe as the Industrial Revolution and mechanisation of production progressed, was a stark problem. Modern agricultural production, based on the application of machines and other innovations, increased yields and freed part of the workforce to do other tasks like working in industry. Additionally, with higher yields came market surplus, which could be placed on the market to be sold or traded. This brought down the prices of agricultural goods.

Croatia remained a backward half-feudal land, lacking significant industrial production, until the mid-nineteenth century. The application of the first steam engine in industry was in 1835 in Rijeka (in the paper industry), and in transport between 1838 and 1840 (steamboats in Sisak and Rijeka). The poor state of industrial development can be seen in data from 1850, according to which Croatia had three steam engines capable of producing 30 horsepower (hp), while Austria had 300 engines (3,500 hp), France more than 5,300 (65,000 hp), and the total strength of steam engines in England exceeded 70,000 hp (Gelo 1987). The application of steam engines in production and the transition from man to machine power in industry was a slow process that only came up to speed in Croatia in the 1860s.

During the Austro-Hungarian era, modern statistical services were used to document statistics in Croatia. This meant that indicators of economic development also began to be documented, enabling historians to see general change in GDP per capita. According to documentation from 1820, in the territory of contemporary Croatia GDP per capita was roughly 513 USD (Stipetić 2003). More significant growth would be documented from 1880 through the start of the First World War, with GDP per capita nearly doubling (Stipetić 2003). This data clearly indicates dynamic economic development, which started the process of industrialisation in Croatia.

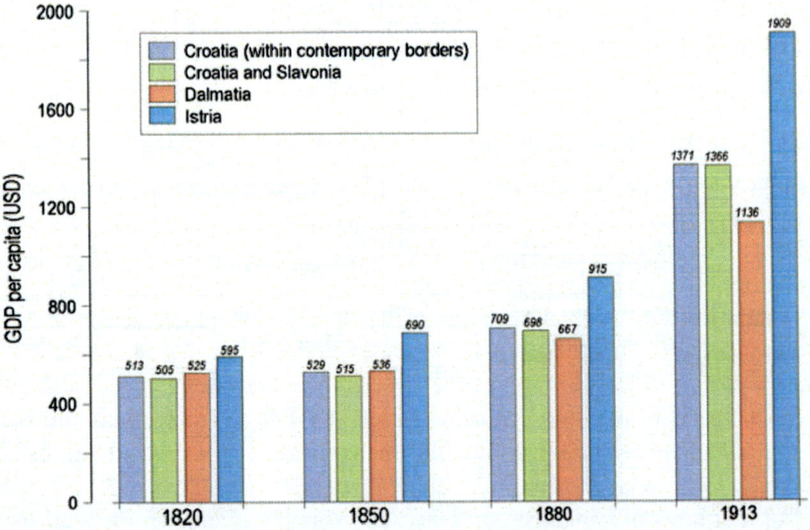

Fig. 8.5 An estimation of GDP per capita change in individual parts of Croatian territory from 1820 to 1913 (After Stipetić 2003, p 146; Adapted from Glamuzina and Fuerst-Bjeliš 2015)

The dynamics of GDP growth were dictated by the Kingdom of Croatia and Slavonia, which encompassed the greater part of contemporary Croatian territory. Therefore, the GDP of Croatia and Slavonia was largely equal to the average for the entirety of Croatian territory (Stipetić 2003).

In the observed era, there was especially interesting GDP change in Dalmatia—higher than the average in other parts of Croatian territory through the mid-nineteenth century, but continuing to lag behind in 1880 (Stipetić 2003). Dalmatia's economy lagged behind (GDP of roughly 83% of the Croatian average) until the start of the First World War.

In contrast, the sharpest GDP growth was found in Istria. Istria had roughly 15% higher GDP in relation to the Croatian average in 1820, and in 1913 this had grown to 40% higher. The difference between Istria and Dalmatia was stark: in 1800, Istria had 15% higher GDP than Dalmatia, and in 1913 this difference had grown to 70% (Stipetić 2003).

Accordingly, the problem of imbalanced economic development of individual parts of Croatia became more prominent during the nineteenth century, along with the beginnings of industrialisation. We can see, from the aforementioned GDP per capita changes, how the process of economic polarisation began during the nineteenth century. This process would result in increasing differences in regional development[9] (Fig. 8.5). This would become more expressed after the fall of Austria–Hungary in the twentieth century, and lead to problems in the development of certain regions that remain unsolved at the time of writing (Stipetić 2003).

8.5 Croatia Proper—Under the Direct Rule of Croatia's Ban

The act of ending feudalism without first undertaking agrarian reforms had negative consequences for rural areas, the most visible of which was the sharp rise in the number of landless peasants. Namely, the advent of a capitalistic renter/landlord paradigm caused the abrupt failure of rural cooperatives, as well as individual plantations owned by nobles that were based on autarchic production and not oriented toward the market. These were converted into small private holdings, upon which agricultural production was expensive and inefficient. As a result, agricultural products from such holdings could not compete on the open market with products from

[9]According to the same research, the index of GDP growth for Croatia (within contemporary borders) from 1820 to 1913 was 267.3, while values for individual regions were as follows: Croatia and Slavonia 270.5; Dalmatia 216.4; and Istria 320.8.

other parts of the Empire or from other European countries. Further fragmentation of small agricultural land parcels resulted in additional population growth, reaching a peak at the end of the nineteenth century. Land was passed on via inheritance and was divided equally among sons and unmarried daughters (married daughters would obtain land via their husbands).

Agricultural production did not pay off on small parcels of land where advanced techniques and technology could not be effectively applied. This resulted in a large number of agricultural workers taking out what amounted to predatory loans, which usually resulted in the workers losing their land. This caused the number of landless peasants to rapidly grow.

The weakly developed industrial sector could only provide work for a small number of landless peasants. Many who could not find work simply emigrated, most often to the Americas. It is important to note how the process of the collapse of cooperatives was strongest in continental Croatia, hitting the eastern parts of the Pannonian Basin and a line along the edge of the former Military Frontier the hardest. In contrast, immigrants from other parts of the Empire (Austria, Czechia, Hungary), who were more experienced with capitalistic renter/landlord relations, came to the area and bought the land of the failed cooperatives. These newcomers practiced agriculture using various advanced technological and technical innovations, including the practice of consolidating smaller holdings into larger ones.

The crisis in Croatian rural areas added to the great European agrarian crisis, which lasted for more than 20 years (1873–1895) and was primarily a consequence of cheap agricultural products from the North American prairies flooding the world market. This was mostly of grains, which dominated the market in Croatia. Grains were farmed on more than two-thirds of agricultural land in Croatia (Bićanić 1975), making the import of cheaper American grain devastating for local agriculture. The losses due to falling prices of agricultural products continued, as did the increase in exploitation of land, forests, and the export of wood (Bićanić 1975). State authorities additionally contributed to the demise of small agricultural workers and the process of deagrarisation by increasing taxes during the highest point of the crisis[10] (Šidak et al. 1968).

The process of breaking apart agricultural holdings is best shown by official data regarding the size of agricultural holdings in Croatia (encompassing nearly all of continental Croatia), including the entirety of Srijem and the Kvarner Littoral.[11] At the end of the great European agrarian crisis (1895), nearly half of all agricultural parcels in Croatian territory (44%) were smaller than 2.8 ha; 27% were between 2.5 and 2.8 ha; 28% from 5.8 to 58 ha; and only 0.26% were larger than 58 ha. In the five-year period that would follow, the number of agricultural parcels would dramatically grow: from 407,000 in 1895 to 520,000 in 1900 (Puljiz 1985). Agrarian overpopulation was a problem under such conditions and was most expressed in Hrvatsko Zagorje and the Kvarner Littoral, where there were nearly 300 inhabitants per km^2.

The penetration and strengthening of capitalist relations in the continental part of Croatia was closely linked to the development of a modern transportation system: in this case, the railway system. Rail connections were a precondition for the process of industrialisation and general economic development, and the level to which they were developed would result in expressed differences in the regional development of Croatian territory. The process of industrialisation would, in any case, be much weaker than in other parts of the Empire (Czechia, Hungary, Austrian Silesia), and mostly took place in the continental parts of Croatia that were linked to the railway network and Rijeka, Croatia's largest harbour.

The number of industrial workers in most industrially developed parts of Croatia at the turn of the nineteenth century to the twentieth century was around 50,000, concentrated in a handful of cities: Zagreb; Varaždin; Karlovac; and Sisak.

[10] In this period (1871–1895), taxes were increased by 500% (measured according to real value).

[11] With the exception of the Međimurje Region in the north, Baranja in the northeast, and Rijeka along with its harbour.

The process of urbanisation was slow to take hold, and near the end of the era of Austrian rule (1910), only 8.5% of the population was located in urban areas (Šidak et al. 1968).

The construction of the railway network also led to neglect of the Pannonian–north Adriatic transportation system and economic stagnation of the areas that depended on its river trade (on the Sava and Kupa rivers). The roads through Gorski Kotar and Lika also fell out of favour.

8.6 The Impact of Political–Geographical Marginality on the Development of Istria and Dalmatia

The formation of a new transport system of railways and dirt roads, as well as maritime routes in the Adriatic, within Austria–Hungary led to the marginalisation of much of the Croatian Littoral. Istria found itself in the best position, with four large harbours in the general vicinity: Trieste; Pula; Venice; and Rijeka. This meant that Istria was on the cutting edge, i.e. a place where lots of new ideas were coming in and lots of things were happening. The greater part of Istria, however, was oriented toward agriculture for income. This meant that the process of deagrarisation was very slow. The share of population involved in agriculture fell by only 11% between 1880 and 1910 (from 72.9 to 61.2%). Commercialisation of agriculture and industrial development took place on the western coast of Istria, while the rest of the area significantly lagged behind in economic terms. Railway and steamboat lines between the western coast of Istria and the eastern coast of the Adriatic also contributed to this state. Individual holdings prevailed in terms of land ownership: 71.2% of land parcels in Istria and on the Kvarner islands were smaller than 5 ha. The bulk of agricultural land in Istria (around 35%) was made up of pastures, and forests made up 22% of the Peninsula (Mirković 1968; Merlić 2008).

The orientation of the majority of trade from domestic to international, via the harbours Trieste and Venice, reflected very poorly on the economic development of Dalmatia. Namely, Dalmatia had no connection to the main railway network of the Empire[12] (Novak 1944). This meant that the region stagnated in terms of overland and maritime trade and the process of industrialisation was very weak—only taking hold at the end of the nineteenth century, and even then only in the largest cities: Split; Šibenik; Zadar; and Dubrovnik. Due to this, the process of urbanisation was limited. The urban population in 1910 was only 10.7%, of which a considerable number still practiced agriculture (Šidak et al. 1968).

The sailboat crisis, which was caused by the advent of steamboats, additionally slowed the development of Dalmatia's main shipbuilding settlements (Mali Lošinj, Dubrovnik, Orebić). This resulted in a strengthened emphasis on agriculture as the primary (often only) means of generating income. The natural conditions of the dominant karst environment and its stingy soil, however, did not offer very good conditions for the development of agriculture. Despite the poor natural conditions for agricultural development, especially crop cultivation, Dalmatia was the most agrarian region of Austria–Hungary.

Dalmatia retained its high percentage of agricultural population through the turn of the nineteenth century to the twentieth century: there was only a slight reduction in the period of 1890 to 1900 (86.1 to 83.8%). According to the census of 1890, the regions with the greatest share of agricultural population were Dalmatia (86.1%); the Kingdom of Croatia and Slavonia (84.6%); Galicia (77.4%); and Bukovina (75.7%) (Šidak et al. 1968).

The advancement of agriculture additionally complicated societal relations that were based on the colonate system—a sort of recidivist feudal system in its own right that was based on peasants paying rent in goods (part of their harvest), rather than with currency. The single new change was that the state's 10% was taken in currency, rather than goods (harvest), from 1827 onward.

[12]Excluding Metković and Dubrovnik, which were connected to the narrow-gauge Bosno-Herzegovinian network.

A monetary payment system (rather than the previous 10% tithe of the harvest) was also in effect on state-owned holdings from 1852 onward. This reform only applied to state lands, which meant that private holdings continued to use the colonate system. Furthermore, there was also forced public service, whereby peasants would be forced to leave the land they worked and go repair roads, and dig or clean canals.[13] The presence of the colonate system slowed the spread of capitalist relations and it was, in the short term, better for the peasantry because it saved Dalmatia from the difficult societal upheaval that afflicted rural parts of continental Croatia.

In contrast, the colonate system prevented the strengthening of commercialised agricultural production. This primarily affected holdings in the vicinity of larger cities, and market production that agriculture in the most developed parts of the world was oriented toward. Additionally, it had a terrible effect on entrepreneurship and incentivising the peasantry to adopt new techniques and technology in agricultural production —iron plows were rare and fertiliser were rarely used.

An additional problem was the small size of land parcels: especially in Dalmatia, where 66.1% of holdings were 5 ha or less, and half of those[14] were smaller than 2 ha. Land parcels on the islands were similarly small, especially on smaller islands. On Šolta, for example, the average land parcel size was 0.05 ha (Peričić 1985). Small land parcels could not generate enough income and, therefore, the capital to improve farming techniques and technology was lacking. The state also had little interest in resolving agriculture problems in Dalmatia, such as irrigation. This can be seen in data from the mid-nineteenth century, according to which one *jutro* (0.58 ha) of land was often chopped up into a hundred tiny parcels. By the mid-nineteenth century, only 53% of the land in Dalmatia was being used for agricultural production, despite significant increases in the total amount of workable land.

The structure of land use, in 1855, was dominated by pastures and forests. It is important to mention here that the quality of the pastures was very low as was forest cover, due to which Dalmatian shepherds would regularly take their herds into Lika and Bosnia to graze. Furthermore, land for crop cultivation (mainly used to grow barley and corn), vineyards, olive and fruit orchards, and personal gardens were very poorly developed[15] (Peričić 1998).

In relation to 1797, the amount of pastures and forests had noticeably shrunk, which indicating a strengthening tendency toward crop cultivation in Dalmatian agriculture. The production structure was dominated by grains, primarily wheat, and vegetables. Grapes and olives were only grown on larger estates. The amount of fields in which potatoes were being grown had increased significantly, largely due to the state handing out seed potatoes for no charge and freeing peasants who grew potatoes from paying their 10% tithe for the first five years. One-fifth of the workable land in Dalmatia was used for growing olives—this was mostly around Šibenik (393,891 trees), Zadar (316,732 trees), and Dubrovnik (307,214 trees). There was also a significant amount of olive cultivation on the Dubrovnik Littoral, the Makarska Littoral, in the surroundings of Split and Trogir, on Pelješac Peninsula, and on the islands Brač and Hvar.

Due to the prevalence of small land parcels and the rising agricultural population, there were frequent famines.[16] By the end of the nineteenth century, despite having the highest percentage of agricultural population of any region in the Empire, the level of production was insufficient to feed Dalmatia's own population—even in rural areas where the food was being produced (Peričić 1980).

[13]It was only in southern Dalmatia that the peasantry worked forced labour for the benefit of the aristocracy rather than the state.

[14]59.1%.

[15]In 1855, there were 720,113 ha of pastures, 275,979 ha of forest, 140,610 ha of farmland, 68,363 ha of vineyards, 15,550 ha of olive groves, and 12,812 ha of orchards and gardens.

[16]Major famines were documented in: 1829–30; 1846–47; 1873; 1877–78; 1879; 1887; 1897; and 1902.

Some much-needed advancement came to Dalmatia, Istria, and the Kvarner islands in the second half of the nineteenth century, thanks to the state of the world wine market at the time. Demand for wine on the European market was going unfulfilled due to the virtual halt of production in Italy caused by the spread of *lug* (*Uncinula necator*), a species of powdery mildew that infected grapevines—for which a cure had not yet been found. Italy significantly increased its import of wine from the Croatian Littoral as a result. This was the era of the first (small) conjuncture (economic boom) that lasted from 1850 to 1857 (when a treatment for powdery mildew was discovered, allowing the Italian wine industry to recover) (Kraljević 1994).

Due to large demand for wine from the Croatian Littoral, primarily Dalmatia, the area used to cultivate olives and grains was reduced to make way for more vineyards (Novak 1944). Vineyards left an indelible mark on the Dalmatian landscape: making up one-fifth of the land on Brač and Vis; one-sixth on Hvar; and one-ninth of the land around Šibenik.

Although a treatment had been found for powdery mildew in Italy, the emergence of a new disease afflicting grapevines—phylloxera (*Viteus vitifoliae Fitch*), which rapidly spread to the eastern Mediterranean and destroyed vast swaths of grapevines—gave a new incentive to build new vineyards. After its wine industry was wrecked by the disease, France began importing large quantities of wine from the Croatian Littoral (1867–1892), which helped the region—Dalmatia in particular—recover economically. This was the period when wine production reached its zenith in Dalmatia and Istria (the great conjuncture, i.e. economic boom). There was a near monoculture level of grapevine cultivation[17] in the area of Šibenik, Zadar, and Split, as well as on the islands of Brač and Hvar (Kraljević 1994), which would later prove to be disastrous. Central Dalmatia led in the share of vineyards in total agricultural area, especially the islands where more than half of the total agricultural land was used for vineyards. The increase in grapevine cultivation was unaffected by the recovery of the wine industry in France, due to the railway and steamboat connections that had been made in the meantime, enabling wine from the Croatian Littoral to be easily placed on markets throughout Austria–Hungary.

The "all eggs in one basket" approach to wine production that took hold on the Croatian Littoral would prove to be rather short-sighted with appearance of the grapevine disease peronospora (*Plasmopara viticola*)—the first cases were documented in Istria in 1880—followed by phylloxera (in Istria in 1884 and in Dalmatia in 1894),[18] in Croatia. This had a similar effect on Croatian wine production to what had happened to Italy and France a few decades earlier: total collapse. This resulted in many vineyards being abandoned. The deathblow to the ailing wine industry in Istria, Dalmatia, and other areas of the Croatian Littoral came in the form of the Wine Clause trade agreement between Austria–Hungary and Italy (1891), whereby the import of Italian wine for low prices was enabled. This began a wave of abandonment of vineyards in Dalmatia and Istria, from which the regions would not recover until after the Austro-Hungarian era. Neither rescinding the Wine Clause in 1904 nor new import taxes levied on the import of foreign wine into Austria–Hungary did much to help the recovery of the Croatian wine industry. Namely, Dalmatia's economy had been broken by disease and the Wine Clause, which resulted in massive waves of emigration. This resulted in the abandonment of vineyards and other agricultural lands—sometimes even entire villages (Kraljević 1994). The consequences of this can still be seen in the cultural landscape at the time of writing.

On the other hand, the lack of a colonate system and dominance of individually owned land parcels in Istria positively reflected on the further development of vineyards. Istrian peasants even managed to weather the vineyard crisis

[17]In the 1890s, vineyards took up 80,000 ha of land in Dalmatia.

[18]The first cases of phylloxera had been documented a few years earlier in the Kingdom of Croatia and Slavonia (1875).

by going into debt. The phase of rapid development of vineyards and the wine industry on the Croatian Littoral had sidelined the spread of orchards—mostly almonds and figs, but to a lesser extent cherries, carob, pears, and apples. The Austrian government incentivised wine production, and the entire amount produced, which was insufficient to fulfil demand, was placed on the local market.

The beginning of the Industrial Revolution in the area of the Adriatic and the wider Mediterranean resulted in the spread of mines, needed to supply raw materials for industrial production. The most important mineral mined during the era was definitively coal, which was extracted along the Croatian Littoral, primarily for the needs of factories in Trieste and Venice. The main coal mines developed on Pag, Brač, in the surroundings of Skradin (Dubravica), and in the area of Drniš (on Promina Mountain as well as in Siverić, where a mine opened in the mid-nineteenth century). Coal mining had already begun in Istria in Raša under French rule, and continued under Austrian rule. Tar was extracted in many locations, and the most important was in Vrgoračka Krajina (Kokorići), which had significant bitumen deposits (Peričić 1998). The construction of railways and harbours also reflected positively on Istrian quarrying. A portion of stone production was meant for export for the needs of Vienna and Budapest, which were expanding intensely during this era.

Factories became new elements in the cultural landscape of coastal cities. These were mostly smaller facilities, e.g. workshops, that made use of local raw materials and produced for the local market. The first manufacturing facilities ran into numerous problems stemming from limited local markets, chronic lack of capital, and competition from quality imported goods, largely from Trieste or Venice. For this reason, many manufacturing facilities closed their doors very soon after opening them.

Until the mid-nineteenth century, the manufacturing facilities that developed in Dalmatia were primarily facilities for fish, candles, soap, ceramics, beer, liquor, flour, olive oil, pasta, textiles and textile dyeing, and leather tanning. Production facilities were mostly located in four large urban centres: Zadar; Šibenik; Split; and Dubrovnik. The opening of cement factories in Split, Kaštela, and Omiš spurred industrial development greatly. Cement production was based on local marl deposits. This also gave impetus for the opening of lime (Siverić, Šibenik, Dugi Rat near Split), and brick (Nin, Solin, Kupari, Imotski) factories. Fish processing took place mainly in island production facilities on Dugi Otok, Brač, Korčula, Hvar, Šipan, and Vis.

Great changes in the landscape were made with the construction of the first hydro-electric power plant. Dalmatia was the pioneer region for this, not only in Croatia but also in Southeast Europe. The first hydro-electric power plant in Croatia was built on the Krka River in 1895. Three more hydro-electric power plants would be built on the Krka River throughout the Austro-Hungarian era, as well as the first such plant on the Cetina River.

Concurrently, fish processing facilities were developed in Istria, as well as facilities for producing tobacco, textiles, wax, soap, alcoholic beverages, and construction materials. The bearers of industrial development were cities on the western coast of the Peninsula, especially Rovinj and Pula.

Weak networks of roads on the Croatian Littoral pushed local agricultural produce to the city markets. Road development, however, was especially dynamic in Istria, where the main road building efforts (on the Littoral) were focused. Dalmatia,[19] on the other hand, was linked to continental Croatia via a road through Velebit, to Bosnia via another road (Split–Sinj–Sarajevo), and to Montenegro (Peričić 1998).

8.7 Industry—the Urban Network

At the beginning of the nineteenth century (1818), long before the start of the railway era, the two largest Croatian cities were Zagreb (population: 8,768) and Varaždin (population:

[19]In 1910, Dalmatia had only 280 km of roads, which was the least of any region in the Austro-Hungarian Empire.

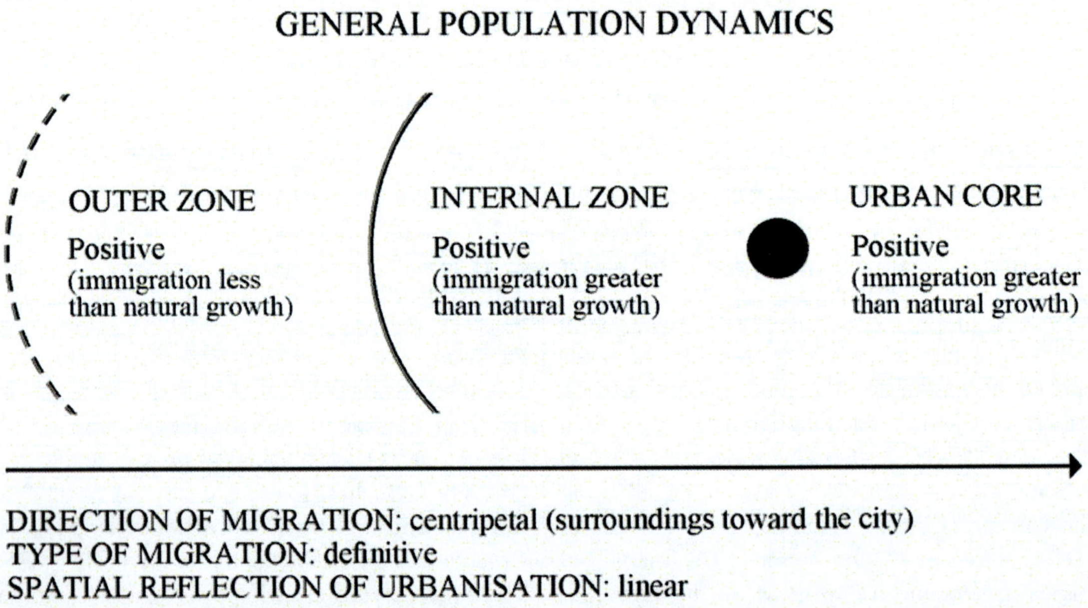

Fig. 8.6 Typical characteristics of the population dynamics of the Zagreb agglomeration during the industrial age (from the turn of the nineteenth century to the twentieth century through the 1950s) (After Fuerst-Bjeliš 1996; Adapted from Glamuzina and Fuerst-Bjeliš 2015)

7,500) (Stipetić 2009). Development of urban networks in the Austro-Hungarian era was based on the principle of industrial development in cities that were linked to the railway network. This allowed the area of northwestern Croatia to assume a nodal role, linking Danubian Pannonia and Central Europe with the northern Adriatic and the greater Mediterranean.

Due to the valorisation of its nodal role, Zagreb was characterised by very dynamic development, reaching a population of 100,000[20] by the start of the First World War. The rapid growth of Zagreb was a result of dynamic economic, primarily industrial, development, which resulted in migration from other parts of Croatia to Zagreb. In the mid-nineteenth century, Zagreb's two constituent settlements, Gradec (Grič) and Kaptol, had grown together into a unified urban settlement (Rogić 1968). The amount of people moving to the city was well documented by the data in the 1910 population census, according to which only 32.27% of Zagreb's residents had been born in Zagreb itself. Moreover, four-fifths of the newcomers came from rural areas, primarily those surrounding Zagreb (Nejašmić 1991).

Construction of the railway line from Budapest to Rijeka, as well as its accompanying link to the *Südbahn*, dictated that the location of industrial facilities. In Zagreb, this meant that most industrial facilities were built along the railway tracks themselves. A direct consequence of this was the linear spatial development of the city from the east to the west (along the railway line). The rapid spatial spread of Zagreb from the turn of the nineteenth century to the twentieth century led to it becoming an urban agglomeration, as it gradually absorbed nearby settlements. The beginning phase of development into an agglomeration was characterised by centripetal role of Zagreb itself: settlements in its surroundings (radius of 10 km) documented 30–50% weaker growth (Fig. 8.6). The creation of a significant number of jobs in the city proper also accelerated depopulation of the outer zones of the expanding agglomeration (Fuerst-Bjeliš 1996).

Population advancement was also documented in other cities, particularly those located

[20]According to the 1910 population census, Zagreb had 100,275 residents.

along magistral railway lines, such that Croatia had 14 settlements with over 10,000 residents in 1910. Karlovac (population: 16,667) enjoyed very dynamic development through the end of the Austro-Hungarian era, according to the census of 1910. It compensated for losing its role as an important river harbour with a railway connection along the Budapest–Rijeka line. Varaždin (population: 13,844), and to a lesser extent Sisak (population: 9,957) and Koprivnica (population: 8,018), similarly enjoyed healthy development up until the start of the First World War.

The railway network also changed the role of certain cities in their area, and increased some of their gravitational areas. A typical example is Graz, which became the main gravitational centre for northwestern Croatia: reaching 50,000 inhabitants. Concurrently, harbour cities strongly gravitated toward Trieste, a railway and steamboat nodal centre that grew into the largest maritime centre in the northern Adriatic. Thanks to its unprecedented trade and maritime significance in the area of continental Croatia, Dalmatia, and Bosnia and Herzegovina, Trieste documented massive urban and demographic development. Trieste dominated the Adriatic in terms of urbanisation; exceeding 100,000 inhabitants by 1860, and its population would grow to 230,000 by the 1914.[21] The influence of Trieste on Istria, Kvarner, and Dalmatia—specifically on cities in those regions—would be a significant development characteristic until the start of the First World War.

In comparison to Trieste, Croatian coastal cities documented humble demographic growth. According to the 1910 population census, Rijeka (population: 67,875) and Pula (population: 59,610) experienced expressed population growth. Split, however, documented much weaker population growth than its northern Adriatic neighbours. The rapid growth of Rijeka and Pula shows the role of state policy and railway connections in urban development. Namely, because it was Hungary's main access point to the Adriatic, Rijeka's population (and harbour) grew far faster than smaller Senj and Bakar. By the mid nineteenth century, after becoming the main military harbour for Austria–Hungary, Pula had transformed from an insignificant urban settlement with less than 4,000 residents to the third-largest Croatian city,[22] thereby taking the role of the main centre of Istria from Poreč.

Dynamic changes in the urban network in Istria are best shown by the data regarding the population of the largest settlements. According to the first official census from 1857, the population of Rovinj was 9,401, Vodnjan (a contemporary suburb of Pula) 3,824, Pula 3,628, and Poreč 2,242. Rovinj would exceed 12,000 inhabitants by the end of the Austro-Hungarian era, while neither Poreč nor Vodnjan would exceed 5,000. In contrast, the main centres in the interior (Pazin, Buje, Buzet) documented only slight growth during the same era (from 2,000 to 3,000).[23]

Development of individual urban centres on the Adriatic coast did not, however, provide the conditions necessary to shape an urban system. This was due to the weak transit connections limiting the strengthening of interactions between smaller cities and the main urban centres (Rogić 1975). Despite lagging behind cities and harbours in the northern Adriatic, Split enjoyed the most dynamic level of development in the central and southern parts of the Littoral. By 1910, Split's population had doubled in relation to the nineteenth century, reaching 32,917. Split led the way in terms of Dalmatia's urban system, in which other large urban centres also documented population growth between 1857 and 1910. Such that the population of Zadar grew from 8,331 to 18,077, Šibenik from 6,611 to 13,659, and Dubrovnik from 7,738 to 12,683. All other settlements had less than 2,000 inhabitants, even those that were linked to the railway and steamboat networks (e.g. Metković, Knin, Makarska).

[21]In the era when the city gained its free imperial harbour status (1719), Trieste had 5,700 inhabitants. When it lost that status in 1891, Trieste had 156,000 inhabitants.

[22]According to the 1910 census, only Zagreb and Rijeka had greater populations.

[23]According to the 1910 census, only Pazin (barely) exceeded 3,000 inhabitants.

Slavonia also documented dynamic development of urban settlements located along the main railway lines, as well as along the Danube River. Osijek took on the role of the main urban centre. It was the most important industrial city located along a magistral railway line. Osijek doubled in population from the mid-nineteenth century to 1910, reaching 34,014 inhabitants, making it one of the largest industrial centres in Croatia. Slavonski Brod (on the Sava River), Vukovar (on the Danube River), and Vinkovci were secondary centres, each with roughly 10,000 inhabitants, while Požega had nearly 6,000 inhabitants.

Political forces would not allow the integration of the three urban networks that had been shaped in Croatian territory (one in the Hungarian part of the Empire, and two in the Austrian part) into a single network. Along with weak transit connections, another cause was very weak economic interaction. Transport isolation dictated the stronger transit and economic connections between Istrian cities, Trieste, and the Slovenian Littoral, while only harbour cities developed in Dalmatia.

8.8 Population Changes—Great Migration Waves

Population changes during this era can be easily reconstructed, thanks to census taking that started in 1857. The average population density of around 30 inhabitants per km^2 and the share of urban population (10%) lagged behind other, more developed parts of Europe. Demographic analyses show a falling birth rate, from 43 to 47% in the peri-Pannonian part of Croatia and 38–42% in Istria and Dalmatia (documented at the start of the nineteenth century) to levels below 30%. A similar fall was also documented in Slavonia, Podravina, and larger urban centres. The falling birth rate indicated the start of the process of demographic transition that began at the end of the nineteenth century, which was also shown by the falling mortality rate. Namely, the mortality rate in Dalmatia and Istria was 25–30% at the end of the nineteenth century, while it was over 30% in continental Croatia. The mortality rate would fall under 25% by the end of the Austro-Hungarian era in all parts of Croatian territory (Gelo 1987; Rogić 1990).

Overseas emigration from Croatia began at the end of the nineteenth century, because the Croatian economy couldn't keep up with population growth, which is characteristic for times of demographic transition. Croatians of the time depended too heavily on agriculture as a way of life and other branches of the economy were not developed enough, so emigration ended up being the "answer" to agrarian overpopulation. Furthermore, Croatian industry was too underdeveloped to absorb the surplus agrarian population as workers, which was not the case in other, more developed parts of Austria–Hungary and Europe (and overseas). The boon of the Industrial Revolution and general economic and social advancement thereof were generally only perceptible in certain urban centres (and even then only in specific areas), and the average inhabitant of Croatia probably didn't notice much of a change in their life.

Emigration was strongest from areas with high agricultural density, primarily those in Croatia's karst and border areas (Dalmatia, Lika, Gorski Kotar, Kordun, Banovina, the Kupa Valley, Žumberak). Emigrants predominantly went to North America, followed by South America. It is estimated that roughly 500,000 people—mostly from Pannonian areas (~350,000), Dalmatia (~100,000), and Istria (~40,000)—left Croatian territory between 1880 and 1914 (Stipetić 2009). A number of emigrants, however, did trickle back to Croatia over time[24] (Nejašmić 1991).

Apart from emigration, there was also immigration into Croatia. Namely, foreign ethnic groups settled in Pannonian Croatia during the nineteenth century, primarily in big cities, where they opened workshops and businesses. There was also an influx of agrarian populations from Germany, Hungary, Czechia, Slovakia, and Ukraine, which left their homes due to economic issues and overpopulation to settle in Pannonia.

[24] In certain areas, 15–30% of emigrants returned to Croatia after many years away.

It is estimated that roughly 150,000 agricultural workers settled in Pannonian Croatia (called Croatia and Slavonia at the time) between 1870 and 1910. The number of people who spoke languages other than Croatian as their mother tongue grew dramatically: the number of German speakers grew from 13,226 (1840) to 134,078 (1910) and Hungarian speakers from 5,050 (1840) to 105,948 (1910). There was also a documented increase in the number of Czechs (31,252 in 1910), Slovaks (21,485 in 1910), Ukrainians (Rusyns) (8,318 in 1910), as well as other groups. Foreign settlers also settled in urban settlements in this part of Croatia.[25]

Immigration of foreigners, primarily Czechs, Austrians, and Hungarians, was also documented in coastal cities; as a result, Rijeka and Pula, and to a lesser extent Zadar and Split, enjoyed stronger economic development (Stipetić 2009). Immigration of Italians, mainly from Apulia, was documented in Dalmatia. They came to Croatia as refugees from the wars of unification in Italy, or as factory and railway workers. They mostly settled in northern Dalmatia (Šibenik, Skradin, Zadar). It is important to state here that migration waves oriented toward coastal cities were significantly weaker and can not be compared to migrations toward Pannonian Croatia.

References

Bićanić R (1975) Ekonomska podloga događaja iz 1903. g. u Hrvatskoj. Historijski zbornik 27–28:49–72

Bunijevac H (2009) Željeznica kao preduvjet gospodarskog razvoja Like. In: Holjevac Ž (ed) Identitet Like: korijeni i razvitak. Institut društvenih znanosti Ivo Pilar – Područni centar Gospić, pp 547–562

Fuerst-Bjeliš B (1996) Zagreb: periodizacija razvoja gradske aglomeracije. Geografski glasnik 58:89–96

Gelo J (1987) Demografske promjene u Hrvatskoj od 1780. do 1981. godine. Globus, Zagreb

[25] An indicator of just how numerous foreign settlers were can be seen in the 1880 census, according to which German was the main spoken language (although, it was not necessarily the mother tongue) of 49.3% of the population of Osijek—the largest Slavonian city.

Glamuzina N, Fuerst-Bjeliš B (2015) Historijska geografija Hrvatske. Sveučilište u Splitu, Split

Iveljić I (2010) Banska Hrvatska i Vojna krajina od prosvjećenog apsolutizma do 1848. godine. Leykam international, Zagreb

Juzbašić Dž (1974) Izgradnja željeznica u Bosni i Hercegovini u svjetlu austrougarske politike od okupacije do kraja Källayeve ere. ANUBiH, djela XLVIII/28, ANUBiH, Sarajevo

Juzbašić Dž (2002) Politika i privreda u Bosni i Hercegovini pod austrougarskom upravom. ANUBiH, posebna izdanja CXVI/35, ANUBiH, Sarajevo

Kraljević R (1994) Vinogradarski slom i demografski rasap južne Hrvatske u osvit 20. stoljeća – vinogradarstvo Dalmacije 1850–1904.: uspon, procvat, klonuće. Književni krug, Split

Merlić E (2008) Buzeština kroz povijest. Vlastita naklada, Pazin–Buzet

Mirković M (1968) Ekonomska historija Jugoslavije. Informator, Zagreb

Nejašmić I (1991) Depopulacija u Hrvatskoj: korijeni, stanje, izgledi. Globus – Institut za migracije i narodnosti Sveučilišta u Zagrebu, Zagreb

Novak G (1944) Prošlost Dalmacije, knjiga druga: od Kandijskoga rata do Rapalskoga ugovora. Hrvatski izdavalački bibliografski zavod, Zagreb

Novak G (1965) Povijest Splita: knjiga treća – od 1797. do 1882. god. Matica hrvatska, Split

Peričić Š (1980) Oskudica i glad u Dalmaciji u XIX i početkom XX stoljeća. Radovi instituta za hrvatsku povijest 13:1–32

Peričić Š (1985) Prilog poznavanju gospodarskih prilika otoka Šolte u 19. st. Geografski glasnik 47:105–120

Peričić Š (1998) Gospodarska povijest Dalmacije od 18. do 20. stoljeća. Matica hrvatska Zadar, Zadar

Population Census in Austria-Hungary (1868–1910), Chech statistical office. www.czso.cz. Accessed 26 Nov 2019

Puljiz V (1985) Socijalno-ekonomska povijest našeg sela. In: Lehpamer M (ed) Zbornik Trećeg programa Radio Zagreba, Zagreb

Regan K (ed) (2003) Hrvatski povijesni atlas. Leksikografski zavod Miroslav Krleža, Zagreb

Rogić V (1968) Historijsko-geografska problematika urbane mreže Jugoslavije u odnosu na osobine njenog suvremenog razvoja. In: Lutovac M (ed) Cvijićev zbornik. SANU, Beograd, p 1968

Rogić V (1975) Osnove historijsko-geografskog razvoja. In: Cvitanović A (ed) Geografija SR Hrvatske – knjiga V: Sjeverno hrvatsko primorje. Školska knjiga, Zagreb, pp 42–45

Rogić V (1982) Historijsko-geografska osnova socijalno-kulturne diferencijacije vojnokrajiškog prostora. Geografski glasnik 44(1):23–38

Rogić V (1990) Regionalna geografija Jugoslavije, knjiga 1 – prirodna osnova i historijska geografija. Školska knjiga, Zagreb

Stipetić V (2003) Stanovništvo i bruto domaći proizvod Hrvatske (1500–1913) u kontekstu najnovijeg rada

Angusa Maddisona. Anali Zavoda za povijesne znanosti HAZU u Dubrovniku 41:91–155

Stipetić V (2009) Stanovništvo Hrvatske u XIX. stoljeću (1800–1914). In: Bratulić J, Vončina J, Jelčić AD (eds) Hrvatska i Europa – kultura, znanost i umjetnost: svezak IV. Moderna hrvatska kultura od preporoda do moderne (XIX. stoljeće). Školska knjiga, Zagreb, pp 13–24

Šidak J, Gross M, Karaman I, Šepić D (1968) Povijest hrvatskog naroda g. 1860–1914. Školska knjiga, Zagreb

9 Croatia Within South-Slavic Political Associations

Abstract

The end of the First World War marked a paradigm shift in politics in certain parts of Europe: primarily Central and Southeast. Following the First World War, old states crumbled and new ones were born throughout Europe. This chapter explores the historical–geographical development of Croatian territory within two South Slavic political entities following the fall of Austria–Hungary: The Yugoslav Monarchy, a continentally-oriented agrarian land; and post-WWII socialist Yugoslavia. Territorial development is explored (banate, oblast, community, and county), along with a review of changing administrative–political divisions throughout the period. After the fall of Austria–Hungary, most of the Croatian territory became part of the newly formed Kingdom of Serbs, Croats, and Slovenes (later the Kingdom of Yugoslavia). The loss of northern Adriatic harbours (Rijeka and Trieste), toward which the most developed areas of the Kingdom gravitated, certainly contributed to the subsequent shift in orientation from the Adriatic to the continent. The period of the Yugoslav Monarchy lasted less than a quarter of a century and was generally characterised by economic and social underdevelopment. Within socialist Yugoslavia, Croatia would see new transport and economic developments, primarily industrialisation and urbanisation followed by deagrarisation and even rural exodus. Zagreb in particular underwent a drastic change, and experienced rapid population growth throughout the whole period, primarily because of its new role as a major industrial and financial centre.

Keywords

Croatia · Kingdom of Yugoslavia · Socialist Yugoslavia · Industry · Urbanisation · Deagrarisation · Zagreb · Banate · Oblast

B. Fuerst-Bjeliš and N. Glamuzina, *The Historical Geography of Croatia*,
Historical Geography and Geosciences, https://doi.org/10.1007/978-3-030-68433-4_9

Vitić skyscraper, Zagreb
An apartment block of the former National Bank of Yugoslavia, built from 1958 to 1962 for all bank employees, according to Yugoslav social policy.
The building gained iconic status as one of the most significant examples of modernist architecture in Croatia. Its nickname is in homage to its architect Ivan Vitić (Photograph: Barbara Bjeliš. Reproduced with permission)

The end of the First World War marked a paradigm shift in politics in certain parts of Europe: primarily Central and Southeast. These changes also had a geographical dimension and made a lasting impression on the historical–geographical development of Croatia. As a result of the tumultuous political–geographical conditions following the First World War, old states crumbled and new ones were born throughout Europe. Croatia was carried along by these changing political and social currents as it passed into a new South-Slavic state from its former position in Austria–Hungary.

9.1 Croatia Within the Yugoslav Monarchy—A Continentally-Oriented Agrarian Land

Although the Yugoslav Monarchy would not last long—less than 25 years—significant social processes were started while it existed. These processes would be accelerated during the second half of the twentieth century under socialist Yugoslavia. The short reign of the Yugoslav Monarchy, as well as numerous internal functional problems, meant that its ability to make a lasting impression in the cultural landscape was limited.

9.1.1 Integration and Disintegration Processes

Croatia was caught up in the sweeping processes of political integration that took place in Southeast Europe at the beginning of the interwar period. The fall of Austria–Hungary in 1918 paved the way for its Slavic-dominated regions to break away and unite politically with Serbia and Montenegro.

Integration was followed by the dissolution of certain peripheral parts of Croatian territory, foremost in the western littoral area. The entirety of Istria and the Cres-Lošinj Archipelago were annexed by Italy. The Free State of Rijeka was also dissolved by Italy after a short period of independence (1921–1924). This denied the newly formed Kingdom of Serbs, Croats, and Slovenes (SHS)[1] access to the northern Adriatic, where the best equipped and most heavily trafficked harbours were located (Fig. 9.1).

Moreover, by annexing some of the smaller, outlying Adriatic islands (Lastovo, Palagruža, and Sušac), Italy was able to monitor all Adriatic transport. Although the Italians did not succeed in annexing the entirety of Dalmatia, as they had sought in the Treaty of London of 1915 (London Pact), they did manage to take Zadar and sever it from its surrounding hinterland. This annexation shows how the Treaty of Rappalo (1920) was regarded as a temporary compromise by Italian leadership, allowing them time to prepare for the conquest of Dalmatia. In contrast to territorial losses on the littoral, the Kingdom of SHS gained territory in the Pannonian Basin, where defeated Hungary had to give up a great deal of territory. This meant that the Međimurje and the southern part of Baranja were ceded to Croatia.

9.1.2 Croatia's Significance Within the Yugoslav State in Terms of Transportation

Although the newly founded South-Slavic state still had a large littoral, it was primarily continentally-oriented. Namely, its inherited transport system, especially the railway network, promoted continental orientation, mostly toward Central Europe. The loss of northern Adriatic harbours (Rijeka and Trieste), toward which the most developed areas of the Kingdom of SHS gravitated, certainly contributed to the shift in orientation from the Adriatic to the continent.

The role of the main harbour was taken over by Sušak, a small port settlement on the left bank

[1]The Kingdom of Serbs, Croats, and Slovenes (SHS) was declared on 12 January 1918 and marked the unification of the State of Slovenes, Croats, and Serbs, the Kingdom of Serbia, and the Kingdom Montenegro. The Kingdom of SHS had a total area of around 250,000 km^2 and roughly 12 million inhabitants. It was only in 1929 that the name was changed to the "Kingdom of Yugoslavia".

Fig. 9.1 Croatian territory within the scope of the Kingdom of Serbs, Croats, and Slovenes (SHS) and its administrative organisation into *oblast* (provinces) (1922–1929) (*Source* Dugački and Regan 2018)

of the Rječina River near Rijeka that was connected to Zagreb by railway. Zagreb had grown to become the main industrial centre of the Kingdom of SHS by this point. The amount of traffic through Rijeka's harbour fell drastically (by nearly 90%) after it came under Italian control, exchanging its role as the main harbour of one state for a peripheral position in another. Citizens of Rijeka could only lament the cruelty of fate as the Kingdom of SHS rerouted the bulk of its goods through Sušak (Rubić 1939; Rogić 1956, 1970a). By the start of the Second World War, the amount of traffic through Rijeka's harbour had fallen to roughly one-quarter (23%) of the amount documented during the Austro-Hungarian era. For Rijeka, being ruled by Fascist Italy—which was not interested in solving any of the city's economic issues—represented a major setback in the city's development (Dukovski 2010).

Due to the insufficient capacity of the harbour at Sušak, development of the harbour at Bakar was also encouraged. This was intended to enable the city to help bear some of the burden of transportation of goods. The Kingdom of SHS also leased part of Rijeka's harbour from the

Italian government to allow them to load and unload goods. A portion of maritime trade also took place through Senj and Karlobag (in the northern Adriatic), as well as through Šibenik and Split in Dalmatia. Some Yugoslav trade still happened via Trieste, due to its advanced infrastructure (the best on the eastern Adriatic coast), but the amount would decrease by half by the end of the 1930s.

Northern Adriatic Croatian harbours, including Rijeka (under Italian occupation), had the largest hinterlands. Nearly 20% of the territory of the Kingdom of SHS gravitated toward them (over 50,000 km^2), upon which lived 30% of the population. Split and Šibenik, in contrast, only became connected to the greater state railway network in 1925, via a new line through Ogulin and Knin (Fig. 9.2). Thanks to this, both cities recorded significant growth in traffic. Split enjoyed the lion's share and was the de facto main harbour of the Yugoslav state until Sušak came to prominence in 1937.[2]

The development of Split's harbour was primarily a consequence of the export of local products. The volume of exports indicated the small size of Split's gravitational area ($\sim$7,500 km^2). Thanks to its new link to the railway and also to the fact that Zadar was no longer competition as a harbour, Šibenik's harbour rose to become the fourth largest in terms of transport volume in the Kingdom. Despite having a larger gravitational area than Split, Šibenik's primary role as naval harbour limited the amount of commercial transport that could pass through it. The loss of Zadar to Italian occupation also resulted in the development of smaller harbours of local significance: Obrovac; Biograd na Moru; and Preko on the island Ugljan.

Dubrovnik's harbour operated under limited conditions, without links to the normal-gauge railway network. Another problem was Dubrovnik's position: far from the economic core of the state in the Pannonian Basin. Despite the aforementioned difficulties, Dubrovnik maintained the third-highest volume of transport in the state until the outbreak of the Second World War. This is primarily a result of its link to the Bosno-Herzegovinian narrow-gauge railway network, which linked it with Bosnia and the interior of Montenegro. The Bosno-Herzegovinian narrow-gauge railway network was connected to the Serbian narrow-gauge railway network in 1925, with the completion of the line from Vardište to Užice. This helped Dubrovnik develop more strongly as a harbour. Additionally, Dubrovnik's Gruž harbour is very deep, and could accept boats with deep hulls. The harbour at Metković in the Neretva Delta, despite having access to Dubrovnik's gravitational hinterland, was of lesser importance as larger boats were not able to sail 20 km up the Neretva River to reach it (Rubić 1939). The construction of a narrow-gauge railway line from Metković to the edge of the Neretva Delta was begun in the 1930s,[3] in order to ensure that Bosnia and Serbia had access to the deep sea trade (with the primary goal of exporting raw materials from the Serbian mines at Bor).

During the 23 years of its existence, the Kingdom of SHS (Yugoslavia) was not able to create conditions suitable for strong development of its Adriatic harbours. This can be seen in the data showing how four-fifths of international trade was done via continental railway or trade along the Danube River. Only one-fifth of international trade was conducted via Adriatic harbours, and most of that was of raw materials such as timber, bauxite, copper, and cement. The lack of interest on the part of the Kingdom of SHS in strengthening its maritime position played right into Italian imperialistic ambitions. Within the scope of policy from the leadership in Belgrade, two magistral railways were built along the Sava River—largely across Croatian territory (see also Fig. 9.2), and the construction of macadam roads was undertaken—though the latter had a very limited impact due to low levels of motor vehicle traffic (Rogić 1970b, 1990).

[2]In 1930, Split was the only harbour city in the Kingdom of Yugoslavia with greater than one million tons of annual traffic of goods.

[3]This line would be finished during the Second World War under the Italian occupation, however, it would not be used due to the lack of harbour infrastructure.

Fig. 9.2 The development of railway networks in Croatian territory within the Kingdom of SHS and the Kingdom of Yugoslavia (1918–1941) (Modified after Glamuzina and Fuerst-Bjeliš 2015)

9.1.3 Economic-Geographical Development

Economic-geographical characteristics of development of the new state show how it was primarily an agrarian land that, in 1939, was on the verge of falling apart. Agriculture made of 55% of GDP, while industry accounted for a mere 15%. This made the Kingdom of Yugoslavia (along with Albania) the least economically developed state in Europe.

The high proportion of the agricultural population was expressed throughout the Kingdom's period of existence. This is well documented by data regarding the practically non-existent process of deagrarisation: the share of agricultural population fell from 78.8% in 1921 to 76.6% in 1931, and further to 75.1% by 1939. The high percentage of agricultural population led to rural overpopulation, which increased from 1921 to 1931. Continual population growth in rural areas was a result of the high birth rate (tending toward a gradual fall throughout the Kingdom's existence), while the high mortality rate (which also gradually fell throughout the Kingdom's existence) was an indicator of the underdeveloped health system and low living standards in rural areas. The high mortality rate was largely a result of the spread of deadly illnesses typical for underdeveloped areas (tuberculosis, typhus, dysentery, malaria).

The growth of the number of peasant holdings, which happened in the period of 1931–1941, show the extensive fragmentation of agricultural holdings into tiny parcels that were unable to provide sufficient food for the needs of the families that lived on them. The majority of the agricultural holdings in this era (67%) had less than 5 ha of land and used backward technology.[4] This meant that they were unable to create a surplus that could be sold on the market. Additionally, agriculture of this era was still based on traditional peasant polycultural production (Mirković 1968).

Increasing agrarian overpopulation and agricultural crises were constants of social development in Yugoslav rural areas, and the worst problems were found in karst and mountainous areas. Agrarian overpopulation was greatest in western karst areas, primarily Dalmatia, where agrarian population density ranged from 250 to 350 inhabitants per km^2 of workable land. In continental Croatia, the most difficult conditions were found in Hrvatsko Zagorje, where the average agrarian population density was 200–250 inhabitants per km^2.

In contrast, the largest agricultural holdings were found in Pannonian and peri-Pannonian areas, and were mainly oriented toward commercial production. The short post-WWI era of increase in agricultural productivity was a result of the political and economic blockade of communist Russia. Prior to the Russian Revolution, the Russian Empire had been one of the largest exporters of agricultural goods. The Revolution and subsequent blockade forced many Central European countries to look elsewhere for agricultural imports that had previously been obtained from Russia. The era of agricultural conjuncture was, however, very short (lasting until 1925) due to the rise of new export powers on the world market (Argentina, USA, Canada).

The Italian annexation of Istria, the Kvarner islands, Zadar, and Lastovo, and their occupation of Rijeka had an expressly negative impact on the economies of these areas of Croatia. This is because at the point when a "firm border" took shape between Italy and the Kingdom of Yugoslavia, the aforementioned areas became a part of the Italian periphery. The amount of vineyards significantly shrank in the cultural landscape of these areas, due to competition from wine produced in other parts of Italy as well as protectionist taxation policies aimed at goods from Istria.

An additional blow to wine production was the introduction of a policy of state-planned agricultural production, under which farmers were forced to cultivate grains, corn, and potatoes—this also hit other polycultures, e.g. vegetables (Antić 1981). The result of this policy was an increase in the area of farm fields and a decrease in the total area[5] of vineyards and olive orchards. This new agrarian policy, however, had weak results and Istria was forced to begin importing grain to satisfy the needs of its own population—which dropped to 23,000 between 1910 and 1927.

The weak industrial sector could not absorb surplus agricultural workers, as there were only 145,000 new jobs created in industry from 1918 to 1938. State support for industrial development showed meagre results. Industrial workers were paid very low wages and there were only 700,000 of them in 1939, which was reflected in the low level of productivity. An additional problem for industrial development was the sluggish advent of electricity. The level of production of electrical energy in 1938 (70 kWh per capita) was well below the European average of 1,000 kWh per capita (Rogić 1990).

Within the Kingdom of SHS (Yugoslavia), there were significant differences in the level of economic development between Slovenia, Croatia, and Voivodina on one hand, and the rest of the state on the other. Economic underdevelopment, poor educational structure, and

[4]In 1920, nearly one-third of the plows in use in the Kingdom were still made of wood.

[5]From 1910 to 1938, the area taken up by olive orchards shrank from 36,268 to 16,754 ha.

organisation into traditional rural households were the dominant characteristics. During its existence, the Kingdom of SHS (Yugoslavia) showed few (if any) signs of change in its inherited social structure. Moreover, the economic structure itself (largely inherited from Austria–Hungary) dictated the orientation of the Kingdom toward Central Europe.

9.1.4 New Developments and Hierarchy Within the Urban Network

Political changes, along with humble economic growth, were reflected in the changes that took place in the network of urban centres in Croatia. The main characteristic of urban development was the abrupt growth of Zagreb as the main industrial and business centre of the Kingdom of SHS—and one of the main centres of the entire state, along with Belgrade. The dynamic development of Zagreb was not enough to stop the new, politically motivated administrative division of Croatia into banates and the territory of the capital, which replaced the old system of oblasts based on historical regions in 1929. The gravitation area of Zagreb—the seat of the new Sava Banate—was reduced by the new administrative division (Fig. 9.3). Zagreb, however, continued to grow economically and in terms of population during the era of the Kingdom of SHS (Yugoslavia), largely thanks to dynamic industrial development as well as valorisation of its transport-geographical location under the new political–geographical conditions.

Namely, Zagreb was able to emerge from under the thumb(s) of Vienna and Budapest—the main centres of economic and political power along the Danube River—with the fall of Austria–Hungary. A new policy of transport connection (primarily via railway) was also created by the fall of Austria–Hungary. The creation of the Kingdom of SHS in turn spurred the political and economic integration of the southern part of the Pannonian Basin, thereby laying the foundation for better transport connections. Valorisation of the Sava River transport corridor was undertaken for this reason, and the far south-eastern part of Europe was thereby better connected with Central and Western Europe. The continual development of the Sava River transport corridor, along with the existing Pannonian–north Adriatic transportation system (which had been heavily valorised during the Austro-Hungarian era), significantly contributed to the development of Zagreb. These changes also turned Zagreb (the largest city in Croatia) into one of the most important transport nodes in Southeast Europe.

The dynamic development of industry,[6] markets, and financial business also played heavily into the development of Zagreb. This can be seen in the growth of the number of employed persons in Zagreb from 1910 (43,871) to 1931 (109,406) of 149%. Zagreb became the strongest industrial centre of the Kingdom of SHS in the second half of the 1920s with some 410 industrial businesses in operation. Concurrently, Zagreb also became the main centre for financial business in the Kingdom with around 200 financial institutions in operation, of which there were 75 banks or credit unions and 38 insurance firms. The first Croatian credit union was opened in Zagreb and was one of the largest lenders for industrial entrepreneurs—larger even than the National Bank of Yugoslavia in Belgrade (Hutinec and Goldstein 2012).

Population growth is the best indicator of the impact of these changes on Zagreb during the interwar period: the city grew from around 100,000 inhabitants in 1918 to over 250,000 by 1939. This growth was primarily the result of intensive migration, largely of the population in Zagreb's immediate vicinity.

In the interwar period, the Zagreb urban agglomeration was still in the stage of industrial development characterised by population concentration within the city itself as well as in

[6] The fall of Austria-Hungary was one factor that spurred the increase of industrial development in Zagreb—the main industrial centres of the former Monarchy were found outside of the Kingdom of SHS. There was a large market within the newly founded kingdom that lacked for industrial products, which gave good incentive for industrial development.

Fig. 9.3 Croatian territory within the Kingdom of Yugoslavia and its administrative organisation into banates (1929–1941) (*Source* Dugački and Regan 2018)

suburban settlements (primarily in the western part of the agglomeration). The expressed population growth dynamic of the city, which would last until the start of the Second World War, was accompanied by an inverse dynamic in its agrarian surroundings. Namely, due to poverty in the agrarian areas around Zagreb, much of its population moved into the city and the peri-urban settlements that extended along the railway in search of the better opportunities that work in industry offered. This led to the integration of these new worker settlements into the fabric of the city and, thus, the strong social and economic transformation happening throughout the greater Zagreb agglomeration (Rogić 1966; Žuljić 1982; Fuerst-Bjeliš 1996).

Osijek, an urban centre that had enjoyed expressly dynamic economic and demographic growth during the Austro-Hungarian era, found itself in an interesting situation at the end of the First World War. With the establishment of the new border in Baranja and Slavonian Podravina[7] Osijek's gravitational area was split between two states. The new border disrupted dike systems

[7]Podravina is a region that lies along the Drava River.

and transport links. Some factories, which had previously imported raw materials from sources now found in another state, found themselves in a tough situation. The border even split some large agricultural holdings in twain.

Concurrently, Split—the seat of the Littoral Banate—was busy distinguishing itself as the second most important centre in Croatia. Split, however, had issues establishing itself as a great urban coastal centre due to the lack of transport links with its hinterland and insufficiently developed harbour infrastructure. Split's marked population growth (from 35,000 to 60,000 over the interwar period) was a result of migration and industrial development.

Rijeka and Zadar began to lag behind as a result of their isolation from their respective hinterlands and immediate surroundings—which was the result of Italian occupation. This can be seen in data regarding population change. Rijeka had, according to the last Austro-Hungarian census in 1910, 67,875 inhabitants—this number only marginally increased to 73,726 by 1931, and fell back to 68,780 by 1948. Over the same time period, Zadar's population dropped from 18,077 in 1910 to 17,358 in 1931, and further to 13,954 in 1948.

During the interwar period, the greatest levels of growth were documented in cities with more than 50,000 inhabitants. Weak functional and infrastructural development of cities with around 10,000 inhabitants was a major problem during this era. The exception to this were cities in eastern Slavonia, which tended toward positive development, while in other parts of Croatia larger urban centres dominated in terms of development. Additionally, cities of 10,000 to 25,000 inhabitants also documented weak development. This was a major problem because cities of this size category were of exceptional importance to the urban network in Croatia.

The Kingdom of Yugoslavia was the least urbanised European state of its day (17.5% urban population in 1931) and showed few signs that it was tending toward urbanisation during its existence. Croatia, however, in relation to the other regions within the Kingdom, documented mild growth in the share of urban population over the same period (from 15 to 18.6%), which is an indicator of the presence of the aforementioned strong development dynamic (Rogić 1977, 1990).

9.2 The Transformation of the Spatial Structure of Croatia Within the Socialist Federal Republic of Yugoslavia (SFRY)

Croatia entered a new historical–geographical era after the end of the Second World War, during which it would undergo a total transformation in nearly every aspect. The marks of many interdependent and intertwined processes are still visible in the contemporary cultural landscape.

9.2.1 Political–Geographical Integration Processes

The basis of a new Croatia was laid during the War of National Liberation via political decision of the State Anti-Fascist Council for the National Liberation of Croatia (ZAVNOH). This decision was enacted while much of Croatia was still under the control of the Independent State of Croatia (*Nezavisna Država Hrvatska* or *NDH*)—a fascist puppet state of Germany and Italy—which itself was divided into Italian and German-occupied zones. Once peace treaties were signed with defeated neighbouring countries after the end of the Second World War, individual federal units of socialist Yugoslavia experienced territorial changes.

Thanks to the victorious National Liberation Movement, Croatia was able to re-annex Istria and the Cres-Lošinj Archipelago, Zadar, the islands of Lastovo and Palagruža, as well as other smaller Adriatic islands that historically belonged to it. The conflict with Italy over Trieste, the largest northern Adriatic harbour—toward which the majority of Pannonia and peri-Pannonia gravitated—was resolved by the creation of a

Fig. 9.4 Free Territory of Trieste (1947–1954) (*Source* Proleksis Encyclopaedia, The Miroslav Krleža Institute of Lexicography, http://proleksis.lzmk.hr. Accessed Nov. 26th, 2019)

special territory. This was the Free Territory of Trieste (1947–1954), which was divided into two military zones: the northern British-American zone (A) and the southern Yugoslavian zone (B). After Italy annexed zone A and Yugoslavia annexed zone B,[8] the Free Territory of Trieste ceased to exist as an independent political entity (Fig. 9.4).

Territorial borders among federal units of Yugoslavia were defined around the end of the Second World War. Thereby, Croatia gained its contemporary borders (the so-called "AVNOJ" borders), which were defined and named for the Anti-Fascist Council for the National Liberation of Yugoslavia (AVNOJ)[9] (Fig. 9.5).

9.2.2 Post-WWII Restoration and Development

Preconditions for development at the end of the Second World War were very weak. Apart from the meagre level of development that the new government inherited, they were also faced with the immediate problem of (re)building the state, and recovering from both demographic losses and material damage. Therefore, a policy of rebuilding, industrialisation, and electrification was enacted within the framework of Yugoslavia's first "Five-Year Plan".[10] One of the major problems was the lack of capital to finance large projects, which was largely compensated for by volunteer work (mostly by youth work initiatives), as well as by burdening agricultural workers with taxes that went to develop industry. The latter led to significant discontent and resistance in rural areas. Other problems that presented themselves in the immediate post-war era were the lack of a properly-trained workforce[11] (for industry), the growth of bureaucracy, and low productivity levels. Nationalisation and confiscation of all economic subjects on the part

[8]A constituent part of zone B was northwestern Istria, which is part of contemporary Croatia.

[9]The Anti-Fascist Council for the National Liberation of Yugoslavia (AVNOJ) would grow into the interim parliament of Yugoslavia at the end of the Second World War. During the second council of AVNOJ in 1943, the decision to make Yugoslavia a federation was made and borders were decided—the exact borders between the federal units would be fixed after the end of the war. These so-called "AVNOJ" borders would persist through the fall of Yugoslavia into the contemporary era.

[10]Five-year plans for economic development were typical for socialist countries that adhered to the Soviet model of a centrally planned economy. Goals for economic development that were to be met were described within the framework of a given five-year plan. Although Yugoslavia applied the Soviet approach to economic development in the beginning, it was quickly liberalised via the application of numerous elements of market-based economics.

[11]The problem of the lack of skilled workers was worsened after Yugoslavia's conflict with (and subsequent expulsion from) Cominform in 1948, which caused many skilled workers to leave for other socialist states.

Fig. 9.5 Croatia within the AVNOJ-determined borders of socialist Yugoslavia (1945–1991) (*Source* Dugački and Regan 2018)

of the state, apart from certain types of workshops, also reflected negatively on entrepreneurship.

Dynamic economic development in Yugoslavia was eventually achieved with financial help from western states (primarily the USA). The bulk of foreign loans were invested in industry, which led to an abrupt economic and social transformation. Serbia had the strongest economy of all Yugoslav republics in terms of GDP, accounting for nearly 40%. Croatia had the second strongest economy, accounting for 30% of Yugoslavia's total GDP.

In terms of development level (expressed in GDP per capita), however, the most developed republic of former Yugoslavia was Slovenia, with a GDP per capita of nearly double the Yugoslav average. Croatia was second in terms of development level, with a GDP per capita nearly 30% higher than the Yugoslav average (Stiperski and Lončar 2008). The bulk of total GDP created in Croatia was generated in the Zagreb macro-region (51%), while Istria and Kvarner stood out in terms of development level and standard of living. The disparities in regional development from the previous century were still visible during the era of socialist Yugoslavia.

These differences within Croatia would become more expressed throughout the socialist era. Istria and Kvarner were as developed as Slovenia, while Lika, Kordun, Banovina, and certain parts of the northern Dalmatian

Hinterland were some of the most weakly developed areas in Yugoslavia (apart from Kosovo, which was by far the least-developed region of former Yugoslavia).

The basis for industrial development was in light industry (primarily metals), the electrical industry, and the chemical industry. The Croatian petrochemical industry was especially prolific, accounting for 40–50% of Yugoslav production of petroleum derivatives and half of the production of artificial fertiliser. By the mid-1980s, Croatia was producing 71% of Yugoslavia's petroleum, 63% of its natural gas, as well as the majority of its sea salt. Classic industry (shipbuilding, metallurgy, and textiles) played a significant role as well. The main bearers of industrial development, along with Zagreb (which was the most significant industrial centre of former Yugoslavia) were Rijeka, Split, Osijek, Varaždin, Vukovar, Slavonski Brod, Karlovac, Pula, and Zadar.

Agricultural production, primarily of grains and industrial plants, developed significantly. Furthermore, Croatia led in maritime activities, tourism, fishing, and harbour transport, in comparison to its fellow member republics within the Yugoslav federation. Looking at the entirety of Yugoslavia in the post-WWII era, Croatia accounted for 22% of total industrial production, 23% of total agricultural production, and 23% of total exports. According to these indicators, Croatia trailed only Serbia (Bertić and Rogić 1987; Rogić 1990).

9.2.3 Population Flows

Demographic renewal in the post-WWII era was followed by numerous changes in the structure and distribution of the population. The high level of population growth in the initial phase was mostly a result of the high birth rate, as well as a steadily falling mortality rate. This rate of growth would slow by the 1970s, when the demographic transition in Croatia wound down.

The economic structure of the population also changed due to the accelerated process of deagrarisation and the transfer of the surplus agricultural worker population to the secondary sector (primarily industry), and later to the tertiary sector (services). These changes were closely linked to expressed migrations from rural areas to urban areas, which caused an exceptionally dynamic process of deruralisation (which later turned into a rural exodus) and urbanisation.

Although the processes of deruralisation, deagrarisation, and urbanisation were present throughout Croatia during the socialist era, there were significant differences in their respective intensities, depending on location. Depopulation in the post-war era was documented in nearly four-fifths of Croatia's settlements. This was primarily the case for rural areas, especially those in karst areas where a dispersed settlement pattern, characterised by a large number of small settlements, dominated historically. These were also areas that were traditionally agrarian and weakly developed. Depopulation also affected the weakly developed urban network—especially weakly developed local and regional centres.

The surplus agricultural population in Croatia after the Second World War was estimated to be roughly 900,000. This surplus was most expressed in Dalmatia, Hrvatsko Zagorje, and in certain parts of Lika, Gorski Kotar, Banovina, and Kordun. The concept of urban-industrial development, in which local centres (municipalities) played a key role, was the main cause of strong deagrarisation and rural exodus. Agriculture took a back seat to industrial development in the estimation of the Yugoslav government. Through the end of the 1960s, this was most evident in Istria and was additionally linked with the emigration of the rural Italian population to Italy, after the border between Italy and Yugoslavia was opened.

Migrations within Yugoslavia during the socialist era largely went from rural to urban settlements, whereby the bulk of the population moved to larger cities—avoiding smaller urban settlements. Roughly 840,000 people abandoned rural Croatia from 1948 to 1981 (on average around 25,000 per year); the most intense migration during this period took place from 1961 to 1981, with around 570,000 residents of Croatia leaving their villages and moving to large cities. An additional problem emerged from the fact that the people leaving villages in droves were largely young people of prime childbearing age, which resulted in further increases in the urban population, population ageing in rural areas, and even "demographic extinction" in certain places (Nejašmić 1991).

Economically motivated emigration (to other states) began at the start of the 1960s. This took the form of temporary "guest worker" employment in foreign countries and also affected rural areas (in Croatia and Yugoslavia as a whole) most heavily. It is estimated that around 325,000 people left Croatia between 1948 and 1991 (the period when the rural exodus was strongest) in search of work in foreign lands (Nejašmić 1991).

9.2.4 The New Urban Network and the Foundation of the Nodal Regional Structure

In the socialist era in Croatia, as well as in other Yugoslav federal units, a new urban network took shape. The slow initial pace of change of the inherited urban network(s) was sped up by strengthening industrialisation and using it as the main motivator for urban development.

The hierarchy of urban settlements was defined by the approach to the planned development of the urban network, as well as of central functions required by urban centres of a certain category:

1. Macro-regional centres with more than 100,000 inhabitants and a corresponding concentration of central functions and institutions of the highest level (macro-regional).
2. Regional centres with more than 50,000 inhabitants and a corresponding concentration of central functions of a high level (regional).
3. Primary sub-regional centres with more than 20,000 inhabitants and a corresponding concentration of central functions of the sub-regional level.
4. Secondary sub-regional centres with more than 10,000 inhabitants and a corresponding concentration of central functions of the sub-regional (second) level.
5. Tertiary sub-regional centres with more than 5,000 inhabitants and a corresponding concentration of central functions of the sub-regional (third) level.
6. Quaternary sub-regional centres with more than 2,000 inhabitants and a corresponding concentration of central functions of the sub-regional (fourth) level.

In keeping with the aforementioned concept, urban settlements were spatially defined into the roles of macro-regional, regional, and sub-regional centres. State administration monitored the concentration of central functions in these urban areas, as well as the corresponding administrative, cultural, and health institutions—depending on the administrative level of the settlement in question (Rogić 1962, 1969, 1973, 1974, 1983a, 1984).

Mid-sized cities (20,000–50,000 inhabitants) documented the most rapid growth, and had more balanced urban networks in their vicinity. This kind of development was also encouraged by federal state elements. The most dynamic development took place in four newly formed macro-regional centres—Zagreb, Split, Rijeka, and Osijek—whereby growth in Zagreb was especially strong (Rogić 1968, 1972, 1978).

Despite the fact that both macro-regional and regional centres had a high concentration of functions—needed for polycentric urban development—Zagreb continued to grow alarmingly fast. Zagreb's dynamic growth and its role as one of the largest and most important industrial and business centres in Yugoslavia, as well as an important transport node, led to the creation of a large number of jobs and the rapid spatial spread of the city (Fig. 9.6).

Population growth was largely concentrated on the western and eastern edges of the Zagreb agglomeration, while the population in the city's

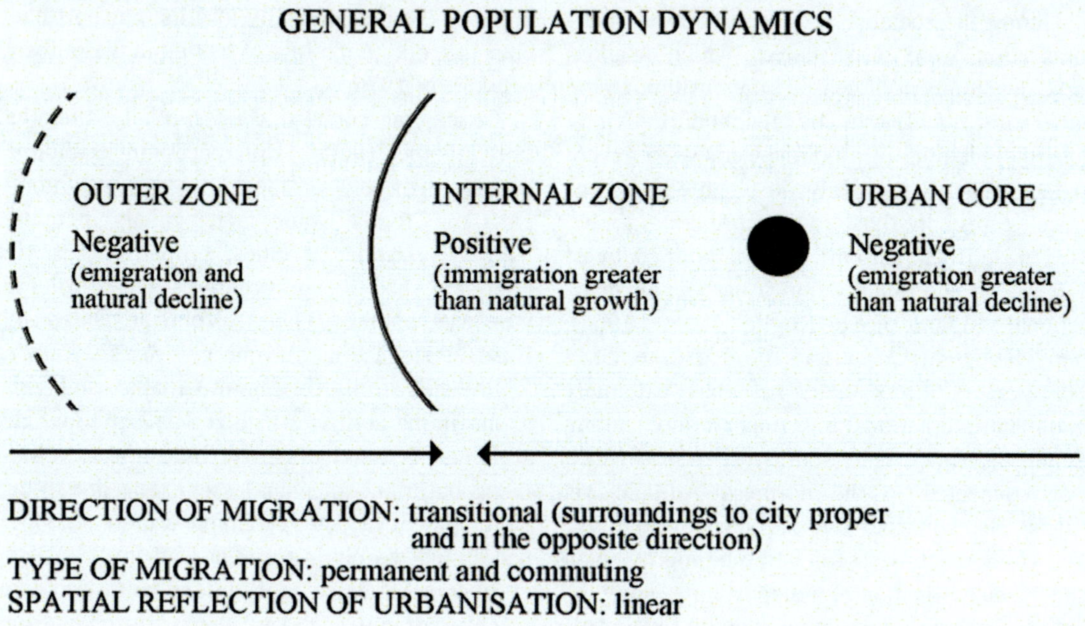

Fig. 9.6 Typical characteristics of the population dynamics of the Zagreb agglomeration during the late (transitional) industrial period (1950–1960) (After Fuerst-Bjeliš 1996; adapted from Glamuzina and Fuerst-Bjeliš 2015)

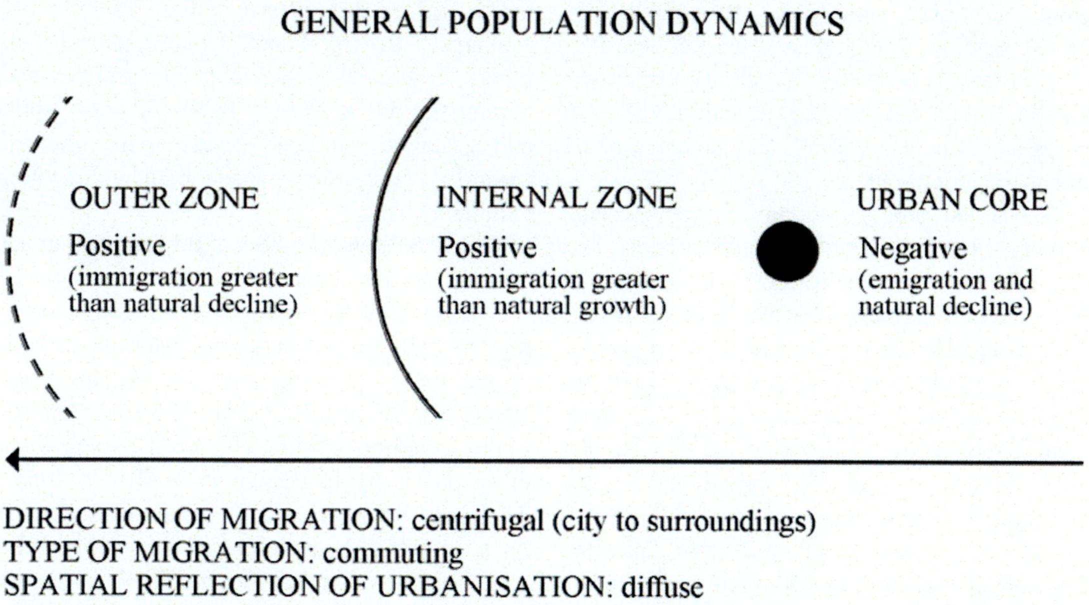

Fig. 9.7 Typical characteristics of the population dynamics of the Zagreb agglomeration during the post-industrial (metropolitan) period (after 1970) (After Fuerst-Bjeliš 1996; adapted from Glamuzina and Fuerst-Bjeliš 2015)

core shrunk The cause of this can be found in the functional definition of the city centre primarily as a business zone, as well as in the consequential movement of the population from the centre to the edges of the city (Fig. 9.7), which is one of the main signifying features of the process of metropolitanisation (Rogić 1983b, 1989; Fuerst-Bjeliš 1996).

During the socialist era, the processes of littoralisation were also started, which sparked rapid development of coastal cities. Although the annexation of Trieste by neighbouring Italy caused something of a political crisis, this event in fact reflected positively on Croatia's harbour cities—especially those in the northern Adriatic. This made Rijeka into the main harbour centre of Yugoslavia and quickly into a harbour of international significance (Rogić 1955, 1964). Improved transport valorisation of the Sava Transport Corridor—part of an extremely important international transport corridor linking Central Europe with Southwest Asia—which was connected to the northern Adriatic via Rijeka, also contributed to harbour development. The continual process of harbour improvement and the electrification of transport links along the railway lines also contributed heavily to Rijeka's development. However, in order to develop into the dominant harbour for an entire state, the organisation of complex harbour infrastructure along the eastern Istrian coast (Raša, Bršica) to eastern Kvarner (Vinodol) and the northern part of the island of Krk was needed. The rejuvenation of Pula's humble port function was also necessary to realise Rijeka's full potential as an international harbour.

During the same period, the southern Adriatic harbour Ploče was built to give Bosnia and Herzegovina more direct access to the Adriatic. Namely, the Neum Corridor was kept as a part of Bosnia and Herzegovina's territory and continued to be the federal republic's only strip of sea coast. The lack of transport links and other coastal settlements in the vicinity[12] meant that Ploče became the most important harbour for Bosnia and Herzegovina.

Transport functions helped develop Split into the leading industrial and harbour centre in the southern Adriatic. Other harbour centres developed at a lower level than Split, e.g. Zadar, Šibenik, Dubrovnik, and (to a lesser extent) the Metković-Ploče conurbation. Dubrovnik, however, would gradually lose its transport harbour role due to the loss of its link to the railway network in 1975.

During the socialist era (especially after the start of the 1960s), differences in economic development between the continental and littoral parts of Croatia would incrementally increase. Namely, continental cities, primarily those in Pannonia, had a significantly stronger role in industry; whereby Osijek, Vukovar, Slavonski Brod, Požega (in Slavonia),[13] Sisak, Kutina, Karlovac, Petrinja (in central Croatia), and Varaždin (in the northwest) stood out. Zagreb would continue its trend of strong industrial development, but to a somewhat lesser extent due to the development of other economic sectors (service) decreasing the city's focus on industry.

In contrast, cities on the littoral belt were more weakly oriented toward industry, with the exception of smaller urban areas in the surroundings of Split (Solin, Kaštela, Trogir). The service industry (tourism and hospitality) dominated in some coastal cities (Dubrovnik, Makarska, Rovinj, Poreč, Umag), transport in others (Rijeka, Ploče), and public services and administration in others (Pula). Šibenik, Split, and Zadar to a lesser extent, were oriented toward multiple economic activities simultaneously.

The accelerated process of urbanisation in the socialist era was most evident in macro-regional centres, and led directly to the formation of urban regions. The Zagreb Region stands out in both spatial and population terms, where there are multiple smaller urban centres (Velika Gorica, Zaprešić, Samobor, Sesvete, and Dugo Selo) in close proximity to Zagreb itself (Rogić 1959). The urban region surrounding Split is the largest on the Littoral, boasting a continual urban zone stretching from Omiš in the south to Trogir in the north. This resulted in the urbanisation of the system of large agricultural fields (Kaštela, Solin, Split, Stobreč), upon which agricultural production developed throughout the twentieth century.

[12]After the construction of the first modern roads (the Adriatic Magistral), a new settlement called Neum grew and gradually became an urban tourist centre.

[13]Požega was officially called Slavonska Požega from 1921 to 1991.

9.2.5 Cultural Landscape Change

The transformation of the cultural landscape was primarily a result of the process of industrialisation, along with the closely related processes of deagrarisation and deruralisation. The change from a largely dispersed settlement pattern, characteristic for autarchic agricultural societies, to population concentration in large urban centres, resulting from the process of industrialisation itself, went hand-in-hand with the changes in the cultural landscape. In this case, the development of urban centres is inseparable from the depopulation and economic stagnation of rural areas.

Namely, the policy of encouraging economic development in urban centres caused large elements of the rural population to migrate to cities, especially to local (municipal), regional, or macro-regional centres. This meant that many hamlets, and even some small rural settlements, were totally abandoned during the 1970s. This was most expressed in the Dalmatian Hinterland, mountainous parts of Croatia, and the interior of Istria. The consequences of these processes, i.e. a staggering amount of abandoned land holdings (so-called "social fallow"[14]), are still visible at the time of writing—primarily in western rural karst areas, which were hit the hardest by push factors.

Deagrarisation was most intense from 1953 to 1971, when the share of agricultural population fell from 56.1 to 32.3%, with an expressed tendency toward further reduction. During this period roughly 40,000 residents of Croatia, i.e. 13,000 families, abandoned agriculture as a primary means of subsistence each year. With regard to this entire period (1948–1981), it is estimated that a total of 1.9 million people abandoned agriculture (57,700 per year).

The cause for this can also be found in the modernisation of agriculture, as well as overpopulation in some rural areas. In contrast, the creation of jobs in other branches of the economy enabled a large number of agricultural workers to find work in non-agricultural sectors. The process of deagrarisation was, however, in many cases not the result of rational calculation, i.e. leaving land when a farmer couldn't make enough produce to make ends meet, but for social reasons. Much of the migration toward cities (and away from agriculture) was the result of the higher perceived social status of non-agricultural workers.

This element, together with existential drivers of migration, pushed people to abandon agriculture in droves—accelerating the process of deagrarisation. This process was so strong that it spun out of the control of state authorities, who were unable to direct it in an organised way (Žuljić 1976). The emergence of social fallow was an indicator of significant economic and social changes in rural areas, caused by the sudden transformation of a largely rural society into an industrial society (Crkvenčić 1981). The industrial sector, which accounted for 22% of total Croatian GDP in 1947, grew to produce 35% by 1982. In contrast, the share of agricultural production in total GDP fell from 38 to 12% over the same period (Gelo 1987).

The era following the Second World War, especially after 1960, was characterised by a slow, but steady, reduction of the area of agricultural land. The amount of land used for vineyards shrank, while the share of pastures and meadows grew, along with orchards. Furthermore, there were significant differences among regions as a consequence of varying environmental conditions and socio-economic development trends.

Farmland prevailed in eastern Croatia, while meadows and pastures dominated along the northern littoral and in Gorski Kotar. Pastures and meadows also dominated in Dalmatia, but there was an above-average share of orchards and vineyards. In northwestern Croatia, there was a healthy mix of all the aforementioned categories. The most dynamic changes in the structure of agriculture happened in the vicinity of large cities, while areas far from larger urban centres (mountainous Croatia and the islands)

[14]This concept was used to describe agricultural land which lay fallow due to economic or societal/social reasons (rather than for the purpose of rejuvenating the soil).

displayed a very slow rate of change (Malić 1983). In contrast, in Pannonian and peri-Pannonian areas, a new agrarian landscape took shape as the area reoriented itself toward commercial agricultural production.

Apart from the physical spread of urban settlements, the largest physical changes in the landscape were caused by the construction of large hydro-electric systems (primarily those on the Drava and Cetina rivers), as well as general hydro-amelioration projects (e.g. in the Neretva River Delta).

So-called social fallow, i.e. unworked agricultural land, was widespread by the 1980s. It had even begun to appear in Pannonian and peri-Pannonian areas—where the largest concentration of the best soil in Croatia was (and still is) located. The causes for this are many and can be classified into four groups: natural (weaker soil, flood plain); economic (lack of interest in agriculture, unprofitability, small area of holdings, distance); social (absentee owner, lack of workers, ageing rural households); and communal (repurposing of land for communal needs) (Crkvenčić 1983). The role of natural factors was key (primarily that of poor soil quality), as well as social factors (population reduction) (Crkvenčić 1982; Malić 1988).

The amount of social fallow spread fastest in karst areas: in Dalmatia (primarily in the Hinterland and on the islands); Lika; and on the Kvarner islands. In these areas, abandonment of agricultural land, especially pastoral land, resulted in the spread of degraded Mediterranean vegetation, i.e. maquis and garrigue (Fuerst-Bjeliš 2018).

A certain type of rural settlement pattern, which population geographers and demographers termed a "depopulation landscape", emerged as a consequence of negative development processes. Apart from abandoned land, a depopulation landscape is also characterised by other abandoned elements of the cultural landscape, such as terraces, stone walls, houses, closed schools, and overgrown gardens (Nejašmić 1991; Fuerst-Bjeliš 2018).

An additional problem in agriculture was the creation of a special social group called worker-peasants. Namely, the people who lived in the vicinity of large cities and had obtained jobs in industry were able to commute to work daily thanks to transportation development. This meant that they could keep their place of residence in their home settlement, and continue to practice agriculture on their land after they came home from work each day. In such cases, agriculture was a source of additional income, or of food for one's own dinner table. This meant that an interesting rural settlement pattern emerged around larger cities, made up of small land parcels where traditional polycultures were grown. While the additional income or wholesome food that some gained from this was beneficial, it also hindered the strengthening and commercialisation of agriculture in these areas (Vresk 1972; Cvitanović and Fuerst-Bjeliš 2018).

Meanwhile, in certain parts of Pannonia (Slavonia, Moslavina, parts of Hrvatsko Zagorje, and the surroundings of Zagreb) market-oriented agriculture developed more strongly. The advancement of agriculture enabled certain agrarian reform measures, aimed at enlarging holdings and creating large agricultural-industrial combines under the direction of the state.

At the beginning of the 1960s, tourism began to develop. This development would become dynamic during the 1980s and was geographically expressed as mass construction of various tourist facilities, lodgings, and the intensive (often overintensive) urbanisation of the littoral zone. Transit connections throughout the greater part of Croatia were solved via the greater Yugoslav transit system. The construction of railways was the first phase, followed by the construction of a modern road network, and the rapid proliferation of automobiles starting in the 1960s. The increasing proliferation of automobiles was a consequence of the reduced importance of the railway system, which intensified when certain local and regional lines were discontinued. This reduction caused certain cities, especially littoral cities (Dubrovnik, and Rovinj and Poreč in Istria), to lose a significant portion of their transport function. Leading up to the 1980s, population concentration and expressed economic development in certain areas painfully highlighted the lack of highways and developed air travel infrastructure (Rogić 1983c; 1990).

A fund for the development of the more weakly developed parts of Croatia was established within the scope of the first post-war economic development plants. This set the foundation of planning and the direction of regional development firmly on a scientific basis, in keeping with official planning policy and new legal regulations. Over time, the number of municipalities in the "insufficiently developed" category was changed via five-year plans, either due to successful development of parts of Croatia or due to changes in the criteria used to categorise levels of municipality development. Banovina, Kordun, Lika, and Dalmatia (all border and peripheral parts of Croatian territory) were numbered among insufficiently developed areas throughout the entire socialist era. The karst hinterland of Zadar (Bukovica) especially stood out in terms of weak economic development and gained special status as a result.

State investment in economic development in these areas bore meagre fruit, while infrastructure investment proved to be a significant aid in raising the local population's standard of living. Within the scope of economic aid for insufficiently developed areas, new factories of certain industrial firms from other parts of Croatia were opened. This helped retain more people in these areas and the process of industrialisation, however, minor in scope, also left its mark. However, the main elements of development plans were often not carried out, either because they were unrealistic, due to mistakes in the appraisals of planning teams, or due to the lack of political will or financial resources necessary for their implementation. There were also cases of extremely poor investment, whereby considerable sums were irrevocably lost (Rogić 1980; Vresk 1990).

9.2.6 Spatial Implications of the New Administrative Organisation

Population and economic changes and the accelerated process of urbanisation during the socialist era demanded new territorial and spatial organisation. The new spatial organisation was in harmony with the policy of federalisation and decentralisation on both on the Yugoslav federal level and the level of its constituent bodies (republics and autonomous provinces). This policy was a reflection of the need for a socio-economic transformation to help solve national questions, such as improving the efficiency of state administration. After multiple administrative and political reforms Croatia was divided into 11 communes of municipalities (*zajednica općina*) with seats in Bjelovar, Gospić, Karlovac, Krapina, Osijek, Rijeka, Sisak, Split, Varaždin, and Zagreb.[15] Decentralisation of the political, economic, cultural, health, and social spheres, from the level of republic government in Zagreb to the level of commune of municipalities, was carried out. Communes of municipalities, as larger spatial units, consisted of a large number of basic territorial units—in this case municipalities (Croatia had 102 municipalities when Yugoslavia collapsed).

Giving great political power to local administration units (in the mid-1970s) provoked the emergence of "municipal monocentrism". In spatial terms, this phenomenon presented itself in the formation of a single urban centre within the municipality, in which all industrial and service functions were concentrated. There would also be a corresponding increase in housing construction. This meant that the municipality centre was also where local political power was concentrated. Such centres had broad authority in directing economic development policy on the local level. Available jobs and housing in municipality centres made for positive population growth. These centres documented dynamic growth and development due to expressed migration tendencies; this was even more pronounced when the municipality centre in question was also a macro-regional centre.

Spatial and urban development was directed via strengthened institutions for urban, local, and

[15]Zagreb was simultaneously the seat of both the Commune of municipalities of the Zagreb area and the Urban commune of municipalities.

regional planning, as well as spatial plans conducted first on the *kotar* (district) level, then on the municipal level starting in the 1960s. These mechanisms, however, failed in many cases, causing the emergence of so-called "wild construction" in many cities. Wild construction refers to the construction of buildings intended for housing (largely family houses) without first obtaining proper permits, which physically and aesthetically wrecked areas on the edges of cities. This phenomenon was strongest in the largest urban centres, but was also very present in local (municipal) centres (Vresk 1990).

The municipality system influenced the development of the Croatian urban system in three key ways: directly—by institutionalising the bulk of state administration in the municipality centre; indirectly—via feedback when influencing individual elements of the social and economic systems; and via the endowment of broad authority to municipality centres to make decisions on the political and economic levels (Toskić and Njegač 2003).

References

Antić V (1981) Istra pod Italijom, Argumenti 4:13–31
Bertić I, Rogić V (1987) Pregled po republikama i pokrajinama: SR Hrvatska. In: Pojatina S (ed) Veliki geografski atlas Jugoslavije. Sveučilišna naklada Liber, Zagreb, pp 175–176
Crkvenčić I (1981) Socijalnogeografski aspekti pojave ugara, odnosno neobrađenih oranica. Geografski glasnik 43:95–107
Crkvenčić I (1982) Pojava ugara i neobrađenih oranica i promjene brojnosti stanovništva SR Hrvatske u posljednjih dvadeset godina. Geografski glasnik 44:3–22
Crkvenčić I (1983) Proces socijalnog raslojavanja sela i pojava neobrađenih oranica i ugara u SR Hrvatskoj (na primjeru triju sela u općini Slavonski Brod). Geografski glasnik 45:43–53
Cvitanović M, Fuerst-Bjeliš B (2018) Marginalization between border and metropolis: drivers of sociospatial change in post-socialist Croatia. In: Pelc S, Koderman M (eds) Nature, tourism and ethnicity as drivers of (de)marginalization: insights to marginality from perspective of sustainability and development. Springer International Publishing AG, Cham, pp 313–327
Dugački V, Regan K (eds) (2018) Hrvatski povijesni atlas. Leksikografski zavod Miroslav Krleža, Zagreb
Dukovski D (2010) Istra i Rijeka u prvoj polovici 20. stoljeća (1918. – 1947.). Leykam international, Zagreb
Fuerst-Bjeliš B (1996) Zagreb: periodizacija razvoja gradske aglomeracije. Geografski glasnik 58:89–96
Fuerst-Bjeliš B (2018) Deforestation and reforestation in Croatian Dinaric karst: human-environmental relations and implications. Ekonomska i ekohistorija 14 (1):136–144
Gelo J (1987) Demografske promjene u Hrvatskoj od 1780. do 1981. godine. Globus, Zagreb
Glamuzina N, Fuerst-Bjeliš B (2015) Historijska geografija Hrvatske. Sveučilište u Splitu, Split
Hutinec G, Goldstein I (2012) Zagreb slijedi Europu. In: Goldstein I (ed) Povijest grada Zagreba: knjiga 2. 20. i 21. stoljeće. Novi liber, Zagreb, pp 2–67
Malić A (1983) Regionalne razlike i promjene površina kategorija iskorištavanja poljoprivrednog zemljišta SR Hrvatske. Geografski glasnik 45:55–73
Malić A (1988) Zemljište i poljoprivreda. Sociologija sela 99–100:77–88
Mirković M (1968) Ekonomska historija Jugoslavije. Informator, Zagreb
Nejašmić I (1991) Depopulacija u Hrvatskoj: korijeni, stanje, izgledi. Globus – Institut za migracije i narodnosti Sveučilišta u Zagrebu, Zagreb
Rogić V (1955) Naše morske luke. Geografski horizont 1 (1–2):11–22
Rogić V (1956) Geografski osnovi razvoja našeg pomorstva. In: Sušić M (ed) Izvještaj o radu IV. kongresa geografa FNR Jugoslavije održanog u Beogradu i NR Srbiji, Beograd, September-October 1955. Savez geografskih društava FNR Jugoslavije, Beograd, p 169
Rogić V (1959) Problem izbora i primjena metoda za određivanje geografskih međa naših gradova. In: Pajković M (ed) Zbornik radova V. kongresa geografa FNR Jugoslavije: NR Crna Gora, Titograd, September 1958. Geografsko društvo NR Crne Gore, Titograd, p 469
Rogić V (1962) Fizionomska i funkcionalna regionalizacija Hrvatske. In: Gams I, Ilešič S, Savnik R (eds) Zbornik VI. kongresa geografov FLRJ: LR Slovenija, September–October 1961. Savez geografskih društava FNRJ – Geografsko društvo Slovenije, Ljubljana, p 279
Rogić V (1964) Geografska regionalizacija primorja. In: Crkvenčić, I (ed) Zbornik VII. kongresa geografa SFRJ: SR Hrvatska, September 1964. Savez geografskih društava SFRJ – Geografsko društvo Hrvatske, Zagreb, p 119
Rogić V (1966) Metropolitanizacija USA i neke karakteristične osobine urbane transformacije u Jugoslaviji na primjeru Zagreba. Geografski glasnik 28:33–58
Rogić V (1968) Prostori općeg makroregionalnog centraliteta Zagreba, Rijeke, Splita i Osijeka u odnosu na regionalizaciju ekonomskog programiranja. In: Kirovski P (ed) Zbornik na VIII. kongres na geografite od SFR: Makedonija, September 1968. Sojuz na geografskite društva na SFRJ – Geografsko društvo na SR Makedonija, Skopje, p 459

Rogić V (1969) Rapska otočna skupina. Geografski glasnik 31:109–124

Rogić V (1970a) Prostor riječke komune: prilog poznavanju regionalno-geografske problematike. Geografski glasnik 32:137–159

Rogić V (1970b) Značenje i razvoj naših glavnih luka. Geografski horizont 16(1–2):1–14

Rogić V (1972) Urbanizacija Jugoslavije. Geografski horizont 18(3–4):19–22

Rogić V (1973) Regionalizacija Jugoslavije. Geografski glasnik 35:13–28

Rogić V (1974) Regionalizacija središnjeg planinskog prostora Jugoslavije. In: Bjelovitić M (ed) Zbornik IX. kongresa geografa Jugoslavije: Bosna i Hercegovina, September 1972. Savez geografskih institucija Jugoslavije – Geografsko društvo Bosne i Hercegovine, Sarajevo, p 409

Rogić V (1977) The changing urban pattern in Yugoslavia. In: Carter FW (ed) An historical geography of the Balkans. Academic Press, London, New York, San Francisco, pp 409–436

Rogić V (1978) Beitraege zur Problematik der Raumgliederung Yugoslawiens. Geographical Papers 4:1–11

Rogić V (1980) Fundamental problems of the geographic regionalisation of Yugoslavia. Geographica Yugoslavica 2:19–29

Rogić V (1983a) Nacrt uvjetno homogene regionalizacije SR Hrvatske. Geografski glasnik 45:75–89

Rogić V (1983b) Zagreb – Verdichtungsraum und Region, Münchener Studien zur Sozial und Wirtschaftgeographie 23:45–52

Rogić V (1983c) Die raumliche Persistenz der Entwicklungsachen Jugoslawiens, Münchener Studien zur Sozial und Wirtschaftsgeographie 23:9–15

Rogić V (1984) Jednostavnost i fleksibilnost koncepta nodalno-funkcionalne diferencijacije SR Hrvatske. Geografski glasnik 46:73–80

Rogić V (1989) A twofold notion of the Zagreb urban region. Geographica Yugoslavica 10:5–17

Rogić V (1990) Regionalna geografija Jugoslavije, knjiga 1 – prirodna osnova i historijska geografija. Školska knjiga, Zagreb

Rubić I (1939) Gravitacijske zone važnijih luka na istočnoj obali Jadrana. Hrvatski geografski glasnik 8-9-10:202–227

Stiperski Z, Lončar J (2008) Changes in level of economic development among the states formed in the area of former Yugoslavia. Hrvatski geografski glasnik 70 (2):5–32

Toskić A, Njegač D (2003) Changes in political and territorial organization and their impact on Croatia's urban system and regional development. Hrvatski geografski glasnik 65(1):7–24

Vresk M (1972) Socijalni ugar i drugi oblici napuštanja agrarne aktivnosti kao posljedica emigracije i socijalnog diferenciranja stanovništva. Geografski glasnik 33–34:79–90

Vresk M (1990) Grad u regionalnom i urbanom planiranju. Školska knjiga, Zagreb

Vujić A (ed) Proleksis encyclopaedia. The Miroslav Krleža Institute of Lexicography. http://proleksis.lzmk.hr. Accessed 26 Nov 2019

Žuljić S (1976) Rast gradskog stanovništva i proces urbanizacije u SR Hrvatskoj. In: Cvitanović A (ed) Centralna naselja i gradovi SR Hrvatske: geografska analiza. Školska knjiga, Zagreb, pp 33–56

Žuljić S (1982) Osnove dugoročnog razvoja Zagreba: svodna studija. Centar za ekonomski razvoj grada – Ekonomski institut – Urbanistički zavod grada, Zagreb

Stepping Out of the Past—Developmental Tendencies of Croatia Since the 1990s

10

Abstract

The recent history of contemporary Croatia represents a step out of the domain of historical geography. This is the period of overwhelming change in every sense. Toward the end of the twentieth century, the political map of the world changed with the collapse of Yugoslavia and the emergence of new states—the Republic of Croatia among them. This was just one part of the whirlwind of change that engulfed the world following the fall of Communism in Eastern Europe. These changes drastically affected the position of Croatia in relation to the former constituent republics of Yugoslavia. The changes took place within Croatia itself, mostly due to the Croatian War of Independence, which had the consequence of creating many spatial, demographic, and economic burdens, as well as changes in relation to other European regions. A period of three decades, however, is too short to define substantial cultural landscape change that could be understood as a kind of a landmark of post-socialist Croatia. Therefore, this chapter aims to tie together the basic processes of the post-socialist and post-war period that have initiated geographically relevant trends. Within the general political–geographical context of change, the chapter explores the geographical dimensions of the War and economic transition, the new administrative–territorial organisation within the framework of the new Croatian state, and the consequent implications for its urban and nodal networks. The trajectories of wider integration processes in Croatia and the implications thereof for the quality of life of its residents are also examined.

Keywords

Croatia · Croatian War of Independence · The Republic of Croatia · Migration · Urban network · Integration processes · Economic transition · Quality of life

Toward the end of the twentieth century, as the final stages of the collapse of Yugoslavia ran their course, changing the political map of the world with the emergence of new states—Croatia among them. This was just one part of the

B. Fuerst-Bjeliš and N. Glamuzina, *The Historical Geography of Croatia*,
Historical Geography and Geosciences, https://doi.org/10.1007/978-3-030-68433-4_10

The Baranja Region, northeast Croatia
The contemporary era is characterised by a total reorientation toward market economy, which can be seen in changes in the cultural landscape. Growth in the number of vineyards is a consequence of the expressed commercialisation in agriculture. In the picture foreground, we see fields used for traditional grain cultivation with vineyards running up Bansko Brdo in the background. (Photograph: Siniša Sakoman. Reproduced with permission)

whirlwind of change that engulfed the world following the fall of communism in Eastern Europe. The resulting transition from socialism to capitalism marked Croatia's entry into a new phase of development. The next few sections will draw attention to the basic processes of the post-socialist and post-war period that initiated geographically relevant trends—in spatial, demographic, economic, and developmental terms.

10.1 The Political–Geographical Context of Change

The new political map of the world was shaped during the last decade of the twentieth century, amid the fall of communist Eastern Bloc states. The polarisation of the world into two blocs (West and East), as such, also ended during the same period. Certain countries, in parallel to the destruction of the political system, began to crumble due to internal issues—one of which was Yugoslavia. Croatia gained independence and international recognition after the fall of Yugoslavia, retaining the borders determined after the Second World War. Later disputes over sea borders with neighbouring states (Bosnia and Herzegovina, Montenegro, and Slovenia) stemmed from the fact that republic sea borders within Yugoslavia had never been precisely defined. Croatia's position on the edge of the Danubian Plain and its access to the Mediterranean Sea via control over the lion's share of the eastern Adriatic coast give it considerable importance. However, Croatia's strategic significance was simultaneously reduced by the changing political conditions that followed the end of the Cold War. The armed conflicts that engulfed the newly independent states that emerged from the fall of Yugoslavia forced Croatia to close its border with Serbia and Montenegro, which severely impacted trade relations. At this point, Croatia reoriented itself toward European Union member states.

10.2 The Geographical Dimension of the Croatian War of Independence

Areas of Croatia, which were occupied by the Yugoslav National Army and paramilitary secessionist units during the Croatian War of Independence (*Domovinski rat*), are still burdened by considerable development problems. The consequences of wartime destruction were multifold and, from a geographical perspective, three things were most affected: the population; cultural landscape; and settlement network.

Demographic consequences can be regarded from a direct or indirect viewpoint. Loss of human life, property, and possessions during war operations numbered among the direct consequences. Population exodus, either to safer parts of Croatia or to foreign states, caused by armed conflict was closely related to the aforementioned. Direct war damages and losses in the areas where fighting took place had a negative impact on settlements, the people living in them, and general demographic development. Furthermore, secondary demographic losses, which were a result of direct wartime losses and damages, were of long-term significance. The main indirect consequences were falling birth rates and changes in the population structure (age, gender, ethnicity, and religion). During this period of heavy demographic losses, a "depopulation landscape" took shape, which was characterised by abandoned homes and farmland as well as general economic neglect. This situation is still expressed today, mostly in border areas, e.g. the Dalmatian Hinterland, Banovina, Kordun, and eastern Lika.

Consequences for the cultural landscape were closely linked to demographic losses. Direct consequences of war operations, i.e. the destruction of physical elements of the cultural landscape, have been largely repaired in the years following the end of the War. The reconstruction of certain economic entities, such as factories that were heavily damaged or destroyed, proved impossible. The inability to (re)start production caused a feedback loop, which reflected negatively on demographic development. Namely, the continuity of the cultural landscape's development was broken, which caused population exodus and long-term changes in the demographic structure. The destruction of industrial facilities meant the loss of many jobs, which forced many people in these areas to try to make a living in another way—usually in jobs in the service sector, or in agriculture (mostly in Pannonia). Moreover, the large share of people who suddenly found themselves unemployed was an ominous indicator of the new economic reality in these areas.

Consequences for settlement networks can be seen in the changes in their functional, gravitational, and hierarchical status, both directly and indirectly. The direct consequences are most easily found in areas that were affected by the battle. Certain rural and urban settlements were either partially or totally destroyed in such areas, which amplified the broader population exodus out of war-afflicted areas. Many of said settlements had a significantly smaller number of residents after the War than before, and a corresponding reduction in economic activity. Furthermore, changes in the character of borders toward other former-Yugoslav states changed the way in which people and goods interacted and moved. This situation was most expressed in the eastern and distant southern parts of Croatia, where transport and economic flows and gravitational processes had to be completely re-directed due to long-term border closure. The aforementioned direct consequences also had, over time, an indirect influence on all aspects of life in war-afflicted areas—especially border areas.

These problems also indirectly affected other parts of Croatia (to a much lesser extent) where there was little or no fighting. Although the War was not enough to disrupt the high level of development in areas sufficiently distant from the front lines. This was partially contributed to by the influx of population from war-afflicted parts of Croatia and neighbouring Bosnia and Herzegovina, which led to the strengthening of the economy in areas of Croatia that were not directly affected by the War, as people and even economic entities flooded into them.

10.3 The Geographical Dimension of the Economic Transition

Around the time that Croatia declared independence, problems associated with the transition began—caused by the change from a single-party system with a state-managed economy to a multiparty democracy with a market-based economy. Other former Eastern Bloc states found themselves in a similar situation when communism fell. The process of transition itself (in all formerly communist states, Croatia included) was accompanied by social upheaval that wrecked a significant amount of economic and social acquisitions. The consequences thereof influenced changes to the cultural landscape. Changes in social conditions, along with a major increase in unemployment and increasing differentiation between social strata, absolutely had an influence on the falling birth rate and general population decrease.

Negative development tendencies were most obvious in border areas of Slavonia, Banovina, Kordun, Lika, and Dalmatia. These areas were hit hardest by fighting during the War. Although the majority of damage has since been repaired, many problems relating to economic, social, and demographic renewal remain to be solved. Despite the major problems and difficulties faced throughout the transition and general recessions at the end of the twentieth century and beginning of the twenty-first century, Croatia is economically much stronger on its own than it was within Yugoslavia—far above the average for former-Yugoslav areas (Stiperski and Lončar 2008).

The modern era is still too young to be able to speak definitively regarding the greater changes that have been taking place in the cultural landscape. However, certain changes were already visible after the first two decades of Croatian independence, foremost in agriculture. Namely, when industrial facilities were closing their doors in large numbers, a large number of jobs were lost. As a consequence, a significant number of households turned to agriculture to make ends meet.

This meant that agricultural land was being used more intensely, which could be observed in the cultural landscape in certain parts of Croatia:

Slavonia; Baranja; Međimurje; Istria; and some parts of Dalmatia. Market mechanisms pushed agricultural producers to adjust their production to market demands, and those in close proximity to large urban centres (Zagreb foremost) were at an advantage. State-owned land that had lain fallow was put back into use and leased out to farmers. Numerous agricultural advancements in agricultural production were also applied—from new cultures to the introduction of crop cultivation in greenhouses. The inherited problem of small-scale land parcelling was still present, however, as well as unresolved problems with irrigation.

At the start of the twenty-first century, the process of deagrarisation continued, which caused social fallow to spread as well. This process was most present in mountainous Croatia, where a bit more than one-fifth (21%) of agricultural land was fallow. This was also the case in littoral regions, foremost in Dalmatia (18%) and along the northern littoral (13.9%).

In contrast, economic diversification in rural areas was observed, primarily along the littoral, thanks to the development of tourism. The commercialisation of agriculture demanded more intense use of agricultural land. This process was most visible in the agrarian landscape of the Pannonian and peri-Pannonian areas of Croatia, especially in the eastern parts of Slavonia and Baranja, where fields and gardens made up 92% of all agricultural land that was in use. The proximity of the large market of the Zagreb urban region positively reflected on the intensification of agriculture in the northwestern part of Croatia, where fields and gardens also led in terms of agricultural land usage (68.3%).

The commercialisation of agriculture was reflected in the growth in crop cultivation in certain parts of the littoral, mostly in the interior of Istria, the hinterland of Zadar (Ravni Kotari), and in specific parts of the Dalmatian Hinterland (Cetinska Krajina and Imotska Krajina). Here, the bulk of agricultural land consisted of pastures (43% in Istria and Kvarner, and 26.6% in Dalmatia). This was most expressed on the Kvarner islands. Dalmatia, however, led Croatia in the share of orchards (21.7%) and vineyards (14.1%) in total agricultural land (Lukić 2012).

Furthermore, the closure of factories also changed the cultural landscape, changing the intended use, function, and physiognomy of the space. New residential and mercantile zones sprouted up where there had once been factories. In coastal cities, tourism facilities increasingly replaced crumbling industry. In the post-socialist era, mercantile centres became symbols of the cultural landscape, much as factories had been a symbol of the cultural landscape during the socialist era. Mercantile centres on the edges of larger cities and those in close proximity particularly stand out in terms of size.

10.4 The Change of Administrative Organisation

Alongside the change in the political system, Croatia ushered in a new system of territorial organisation, whereby the state was divided into 20 counties[1] and the City of Zagreb (which was given special status and privileges) (Fig. 10.1). The old system consisting of 102 municipalities was discarded and the territory was chopped up much more finely by 2010: into 556 cities and municipalities. The majority of these newly created administrative units, however, encountered a plethora of financial and functional problems very soon after their formation.

The new administrative structure was decidedly convoluted, and its formation was not adequately based on spatial organisation research. Later corrections to certain county borders, as well as functional problems of certain counties, were also a result of this. Another issue was that not all county centres had sufficiently strong gravitational strength for the administrative role that had been given to them. Due to this, the need for decentralisation and a reduction in the number of administrative units quickly emerged.

[1]The newly formed territorial units were referred to using the historical term for the territorial units in Croatia that had been used from the Middle Ages until the First World War: *županija* (county). This was chosen for the sake of historical continuity, and it is important to stress that contemporary županijas do not cover the same area, nor do they have the same political position, as their namesakes. The administrative seats are also often not in the same cities.

Fig. 10.1 Administrative–territorial division of Croatia into counties starting in 1997 (*Source* Statistical yearbook of the Republic of Croatia 2018, Croatian Bureau of Statistics)

10.5 Urban Network Development Tendencies and Change in Gravitational and Nodal Relations

The consequences of war operations and destruction, as well as changes in the economic system, led to changes in the urban network. Wartime conditions of political and economic centralisation additionally spurred the development of Zagreb, while relatively weakening the significance of Croatia's other three macro-regional centres (Osijek, Rijeka, and Split). The most intensive socio-economic transformation took place in settlements which were part of large urban agglomerations. This process took place first and with the most intensity in the area of Zagreb.

The City of Zagreb—Croatia's largest city and main urban centre—has a special status in formal, administrative, and institutional terms. According to the population census of 2011, Zagreb had

Fig. 10.2 Administrative area of the City of Zagreb within Zagreb County (*Source* Statistical yearbook of the Republic of Croatia 2018, Croatian Bureau of Statistics)

790,017 residents. Its total area within the framework of its administrative boundaries[2] encompasses 641,32 km^2 (Fig. 10.2).

The Zagreb peri-urban area[3]—a dynamic, transformational peripheral zone—lies outside of these administrative borders. According to the census of 1931, pre-WWII Zagreb had 185,581 residents, living in a significantly smaller area.[4]

[2]The contemporary administrative area of the City of Zagreb encompasses a spacious area, stretching from the peak of Medvednica Mountain in the north to the settlements of Dragonožec and Kupinečki Kraljevec (in the Turopolje Region) in the south, and from Podsused in the west to Dugo Selo in the east.

[3]The process of peri-urbanisation is understood to include the development of settlements in the immediate area of larger cities in order to assume some of the larger city's population and a portion of its central functions. Certain authors use the term suburbanisation as a synonym for peri-urbanisation, however, this is not ubiquitous. The fundamental difference in the use of these two terms is how, with the process of suburbanisation, the transformation of the urban periphery is regarded primarily through the relationship of the main city and its satellites, while with peri-urbanisation the focus is on spatial change in the surroundings of the city.

[4]The area from the contemporary neighbourhood Maksimir in the east to the neighbourhood Černomerec in the west, and from the neighbourhood Mihaljevac in the north to the Sava River in the south.

The process of urbanisation has steadily expanded the borders of the City of Zagreb throughout the twentieth and twenty-first centuries, and transformed the nearby rural landscape into a new type of mixed peri-urban cultural landscape. The process of transformation had mainly diffuse spatial characteristics (Fuerst-Bjeliš 1996). With its peri-urban ring, Zagreb has 1,107,622 residents (according to the 2011 census).

The oldest, northern sector of Zagreb's peri-urban area became part of Krapina-Zagorje County under the new county organisation. This, however, did not weaken the process of peri-urban development in this area, where the linear conurbation of Zaprešić–Zabok–Krapina formed. The southern part of the peri-urban sector has enjoyed a significant development dynamic, wherein we find the unique example of Velika Gorica—a transformational incubator that grew from a humble local centre in Turopolje (with less than 2,000 inhabitants in 1931) to 5,000 by 1948, and on to exceed 34,000 inhabitants according to the 2011 census (Vresk 1997).

An exceptional imbalance in city size emerged in the urban network in northwestern Croatia, largely due to the overwhelming concentration of population and economic activity in Zagreb. Other large urban centres in this part of Croatia (Karlovac, Varaždin, Sisak) have around 50,000 inhabitants. In the northern littoral, including Kvarner and Istria, the situation was similar. Apart from Rijeka (~200,000 inhabitants) and Pula (~60,000 inhabitants), there are no other large(r) urban centres (Toskić and Njegač 2003).

Due to the collapse of Yugoslavia, and changes in the character of borders and spatial relationships, the gravitational influence of certain cities was reduced (Dubrovnik, Osijek, Slavonski Brod, Split). The cities that were affected by this were mostly those with gravitational areas that extended into the territory of neighbouring former-Yugoslav federal republics/territories, e.g. Bosnia and Herzegovina, Voivodina, and Montenegro.

Wartime destruction caused during the Croatian War of Independence, the depopulation of rural areas, and economic problems caused by the transition process also all negatively influenced the network of central settlements. The most dense network of central settlements today is in the far north and northwestern parts of Croatia (Hrvatsko Zagorje, Međimurje, Podravina). In contrast, the least dense network of central settlements is found in the border regions of Lika, Gorski Kotar, as well as parts of Kordun, Banovina, and Pokuplje (Lukić 2012).

Vukovar (Slavonia), Karlovac (Kordun), Sisak (Banovina), Gospić (Lika), Šibenik, and Zadar (Dalmatia) are the best examples of town centres which especially suffered and lost significance due to direct and indirect consequences of wartime damage and depopulation of their former gravitational areas. Mountainous Croatia presented a unique problem, as there was no single centre that had developed to the point of assuming the role of the main centre for the area.

The lack of strong regional centres (cities with 50,000–100,000 inhabitants), which could assume the role of bearers of regional development in the urban network, can be clearly seen (Toskić and Njegač 2003). The causes of this are multi-faceted, and their roots can be traced to the administrative–territorial division from the socialist era, which did not encourage the growth and development of mid-sized cities. The development of a larger number of mid-sized cities, which assumed the role of regional or sub-regional centre, only took place in the southern and eastern parts of the country. In the southern part of the country (Dalmatia), alongside the macro-regional centre Split, important roles were played by the regional centres Zadar, Šibenik, and Dubrovnik. In the eastern part of Croatia (which includes central and eastern Slavonia and Baranja), the regional centre Slavonski Brod and the sub-regional centres Vinkovci, Vukovar, Požega, and Đakovo were of significant importance, alongside the macro-regional centre Osijek.

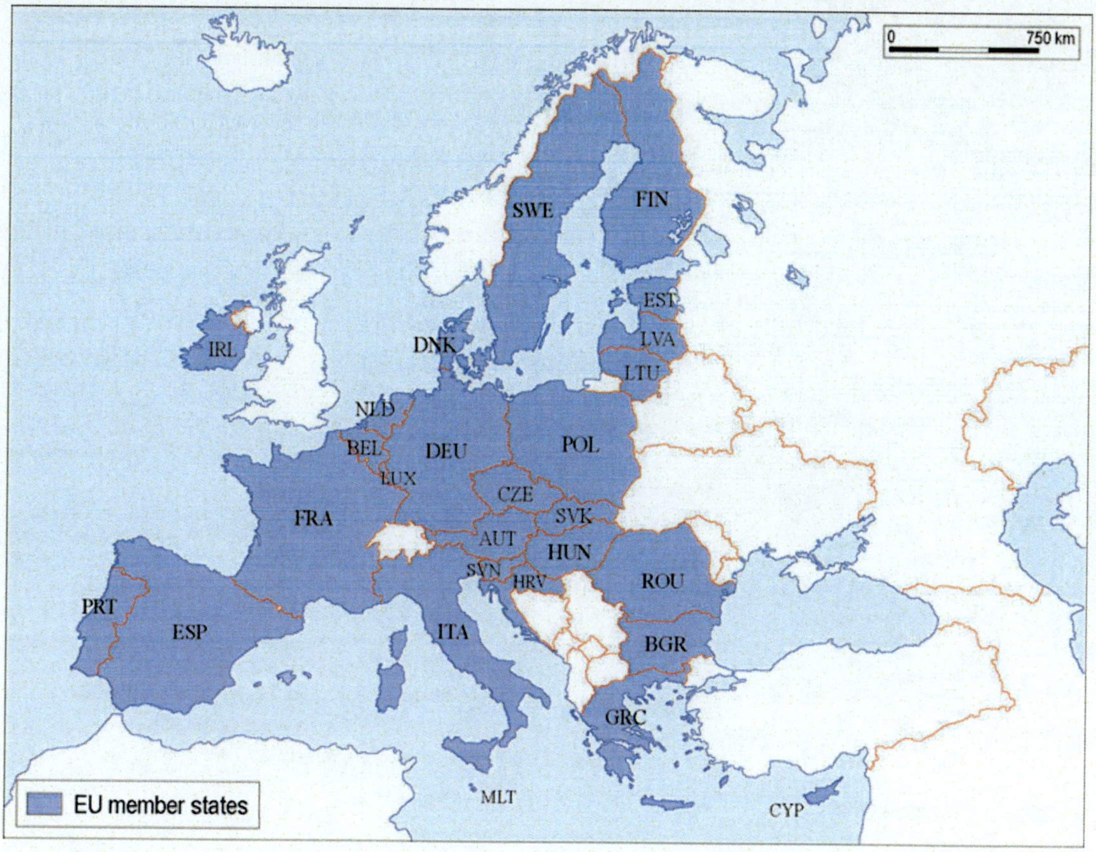

Fig. 10.3 Croatia within the European Union from 1 July 2013 (*Source* EU Open Data Portal 2020)

10.6 Integration Processes and Their Implications for Economic Development and Quality of Life

With Croatia's accession to the European Union in 2013 (Fig. 10.3) and entry into NATO, the process of Euro-Atlantic integration came to a close (Fig. 10.4). Furthermore, with the founding of the Union for the Mediterranean in 2008, Croatia became a part of a multilateral partnership of EU member states and other Mediterranean countries as part of the process of Euro-Mediterranean integration (Fig. 10.5). Croatia, as a Mediterranean country and EU member state, thereby gained the ability to better solve problems relating to economic cooperation and maritime conservation.

The influence of internal disintegration processes within former-Yugoslavia on one hand and the Croatian War of Independence and European integration processes on the other changed the economic position of Croatia within the region in a broad sense. Although Croatia and Slovenia had been the most developed and strongest republics within Yugoslavia in economic terms, Yugoslavia's collapse and the subsequent military conflicts that followed it led to significant changes in the relations between the three largest economies of the former state. Croatia and Slovenia experienced significant economic growth, while Serbia, and the other former-Yugoslavian states, went into recession and economic crisis. Despite the War, difficulties with the transition, and general crisis and recession, the position of Croatia was strengthened, and it continued to distance itself from the

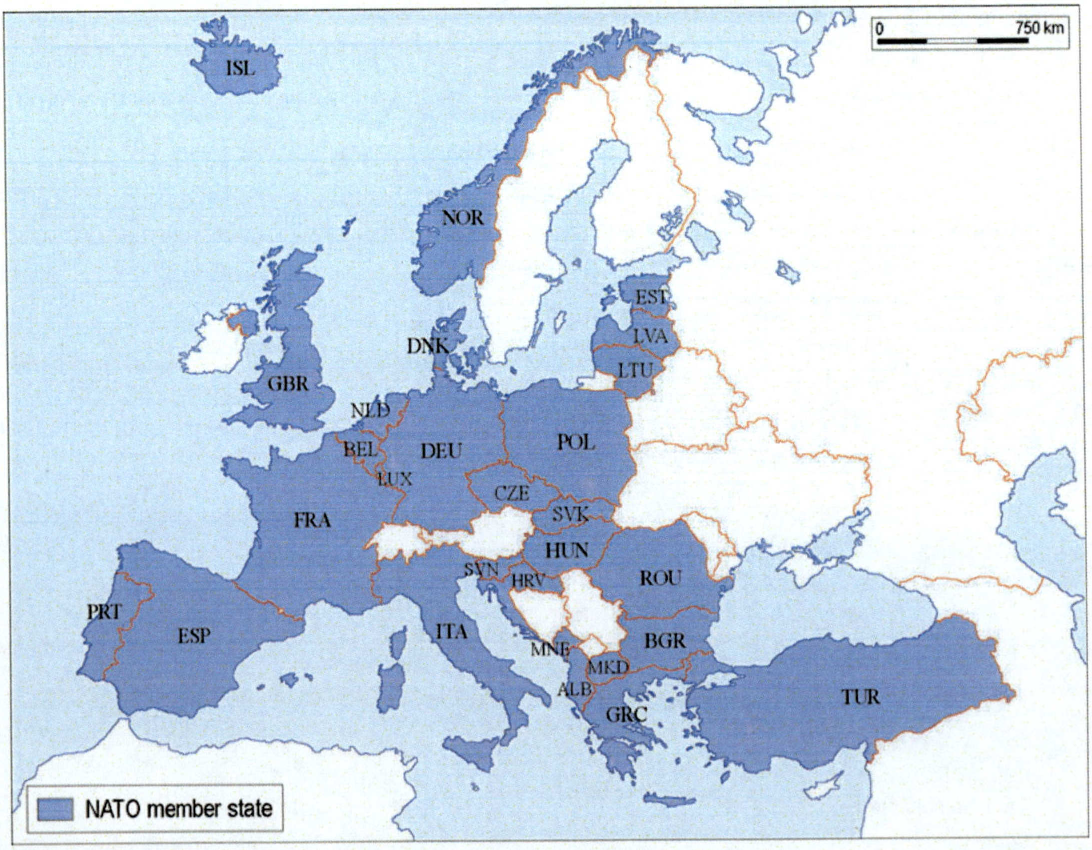

Fig. 10.4 Croatia—NATO member state (*Source* NATO 2019)

average levels of development among former-Yugoslavian states. The disparity in development level among the states that emerged from Yugoslavia became unusually large over a very short period of time (Stiperski and Lončar 2008).

Indicators of standards of living and quality of life in former-Yugoslavian countries, from the start of the process of EU integration, show that Croatia was in the more-developed category along with Slovenia and Hungary. The fact that Croatia took nine years longer than Slovenia and Hungary to enter the EU (in 2013 as opposed to 2004) was largely due to consequences stemming from the War, which its northern neighbours were not burdened by.

According to the Purchasing Power Parity (PPP)[5] indicator, which best reflects the real standard of living, Croatia, Slovenia, Hungary, and recently Romania, significantly distanced themselves from other Southeast European countries (especially other former-Yugoslavian countries—which still find themselves below the world average) over the period of 2008–2018 (Table 10.1).

Croatia is also among the most developed countries of Southeast Europe, along with Hungary and Slovenia, according to the Human Development Index (HDI), which takes PPP into account along with other relevant indicators such as education and average lifespan. According to the 2019 UN Human Development Report (UNDP), a bit less than one-third of the world's

[5]Purchasing power parity (PPP) is an economic theory that compares different countries' currencies through a "basket of goods" approach. It is a theoretical exchange rate that compares the relative cost of buying the same amount of goods and services by country. It is clearly determined by the relative cost of living and inflation rates in different countries.

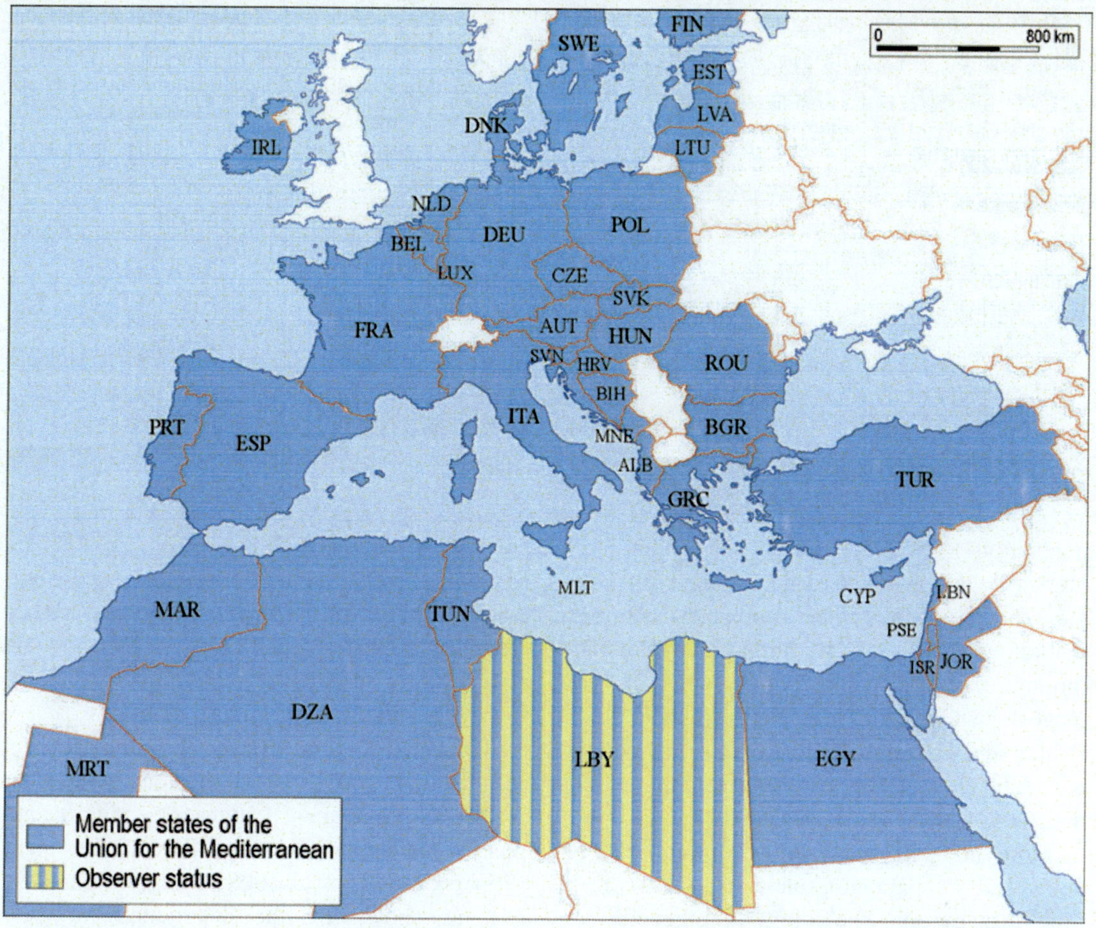

Fig. 10.5 Croatia—member state of the Union for the Mediterranean (*Source* Union for the Mediterranean, www.ufmsecretariat.org. Accessed 20 June 2019)

Table 10.1 Purchasing Power Parity (PPP) per capita of individual states of Southeast Europe from 2008 to 2018

Country (year of EU accession)	PPP 2008 (in USD)	PPP 2013 (in USD)	PPP 2018 (in USD)
Hungary (2004)	19,790	20,930	27,144
Slovenia (2004)	27,870	27,680	32,143
Bulgaria (2007)	11,790	15,200	19,646
Romania (2007)	13,450	18,060	23,906
Croatia (2013)	17,660	20,370	23,061
Bosnia and Herzegovina	8,100	9,820	12,690
Montenegro	13,390	14,600	17,511
North Macedonia	9,340	11,520	12,874
Serbia	10,550	12,020	15,218
World (average)	10,342	14,210	15,745

Source Der Fischer Weltalmanach 2011, 2015, 2019; World Bank; Human Development Reports 2014, 2019 UNDP

Table 10.2 Change in ranking of countries in Southeast Europe according to the Human Development Index (HDI) from 2008 to 2018

Country (year of EU accession)	HDI rank in 2008	HDI rank in 2013	HDI rank in 2018
Hungary (2004)	38	43	43
Slovenia (2004)	26	25	24
Bulgaria (2007)	56	58	52
Romania (2007)	62	54	52
Croatia (2013)	45	47	46
Bosnia and Herzegovina	75	86	75
Montenegro	64	51	52
North Macedonia	68	84	82
Serbia	65	77	63

Source Der Fischer Weltalmanach 2011; 2015; 2019; World Bank; Human Development Reports 2014; 2019 UNDP

population lives in countries marked by high human development. Croatia, along with Hungary and Slovenia, number among the countries marked by very high human development (around the 50th rank), while other former-Yugoslavian countries (e.g. Serbia, and foremost Bosnia and Herzegovina and North Macedonia) are defined as European developing regions according to their HDI (Table 10.2).

More complex analyses of quality of life[6] have clearly separated Croatia from the core of Southeast Europe (including all other former-Yugoslavian states apart from Slovenia). Namely, Croatia, along with Hungary and Slovenia, enjoy quality of life on a level closer to other EU states, i.e. countries with middle-income economies, than to the countries that make up the core of Southeast Europe—which have significantly lower quality of life (Fuerst-Bjeliš and Stiperski 2006.).

The development dynamics of Croatia and other countries in Southeast Europe show that the most weakly developed countries in the region (those that once made up Yugoslavia and are still outside the EU) have similar trends in terms of economic indicators (PPP) to those that have been EU member states for years, but with smaller values. However, differences can be seen in development indicator trends, relating to more complex quality of life (HDI): Croatia and Slovenia have a stable and insignificantly fluctuating rating, while nearly all other former-Yugoslavian states (apart from Montenegro and only recently Serbia) show a temporary tendency toward fall in terms of quality of life.

The small internal market of Croatia, as well as the War and political tensions with other former-Yugoslavian countries, significantly pushed market flows toward EU member states (Stiperski and Lončar 2008). Integration into the EU (as shown by the experiences of other member states) puts Croatian regions, which border on EU member states, in a much better position than other parts of the country. Therefore, it can be expected that these parts of Croatia will document stronger economic growth as a result.

10.7 Closure

The recent period of contemporary Croatia represents a kind of a step out of the domain of historical geography. Therefore, the chapter aimed to tie together the basic processes of the post-socialist and post-war period that initiated geographically relevant trends.

Changes that occurred with the collapse of Yugoslavia drastically affected the position of Croatia in relation to the former constituent republics of Yugoslavia. The changes took place

[6]The Quality of Life Index includes a large number of indicators related to the economy, education, health care, housing, the environment, and socio-political conditions.

within Croatia itself, mostly due to the Croatian War of Independence, which had the consequence of creating many spatial, demographic, and economic burdens, as well as changes in relation to other European regions.

Within the general political–geographical context of change, this last chapter of the book discussed the geographical dimensions of the War and economic transition, the new administrative–territorial organisation within the framework of the new Croatian state and the subsequent implications for its changing regional, urban and nodal networks. The trajectories of wider integration processes in Croatia had the implications thereof for the quality of life, placing Croatia close to central European countries.

A period of just three, although very turbulent, decades is too short to define substantial cultural landscape change that could be understood as a kind of a landmark of post-socialist Croatia. We can only recognise and trace some of the possible paths of geographical development of contemporary Croatia into the future.

This is the end of *our* story but, of course, the story of Croatia will continue. Time goes on, and with every moment it adds one more brick to the road of historical geography.

References

EU Open Data Portal (2020) www.data.europa.eu. Accessed 1 February 2020

Fuerst-Bjeliš B (1996) Zagreb: periodizacija razvoja gradske aglomeracije. Geografski glasnik 58:89–96

Fuerst-Bjeliš B, Stiperski Z (2006) South Eastern Europe: Discussing quality of life indicators, In: Henkel R (ed) South Eastern European countries on their way to Europe: geographical aspects. IFL, heft 5, Leibniz Institut Für Länderkunde, Leipzig, pp 23–38

Lukić A (2012) Mozaik izvan grada: tipologija ruralnih i urbaniziranih naselja Hrvatske. Meridijani, Samobor

NATO (2019) www.nato.int. Accessed 20 June 2019

Stiperski Z, Lončar J (2008) Changes in level of economic development among the states formed in the area of former Yugoslavia. Hrvatski geografski glasnik 70 (2):5–32

Statistical yearbook of the Republic of Croatia (2018) Croatian bureau of statistics, Zagreb

Toskić A, Njegač D (2003) Changes in political and territorial organization and their impact on Croatia's urban system and regional development. Hrvatski geografski glasnik 65(1):7–24

Union for the Mediterranean (2019) www.ufmsecretariat.org. Accessed 20 June 2019

Vresk M (1997) Suburbanizacija Zagreba. Hrvatski geografski glasnik 59(1):49–70